Sanaa Salama

**Antenna Design Challenges on Small Platforms**

Sanaa Salama

# Antenna Design Challenges on Small Platforms

Südwestdeutscher Verlag für Hochschulschriften

**Impressum / Imprint**
Bibliografische Information der Deutschen Nationalbibliothek: Die Deutsche Nationalbibliothek verzeichnet diese Publikation in der Deutschen Nationalbibliografie; detaillierte bibliografische Daten sind im Internet über http://dnb.d-nb.de abrufbar.
Alle in diesem Buch genannten Marken und Produktnamen unterliegen warenzeichen-, marken- oder patentrechtlichem Schutz bzw. sind Warenzeichen oder eingetragene Warenzeichen der jeweiligen Inhaber. Die Wiedergabe von Marken, Produktnamen, Gebrauchsnamen, Handelsnamen, Warenbezeichnungen u.s.w. in diesem Werk berechtigt auch ohne besondere Kennzeichnung nicht zu der Annahme, dass solche Namen im Sinne der Warenzeichen- und Markenschutzgesetzgebung als frei zu betrachten wären und daher von jedermann benutzt werden dürften.

Bibliographic information published by the Deutsche Nationalbibliothek: The Deutsche Nationalbibliothek lists this publication in the Deutsche Nationalbibliografie; detailed bibliographic data are available in the Internet at http://dnb.d-nb.de.
Any brand names and product names mentioned in this book are subject to trademark, brand or patent protection and are trademarks or registered trademarks of their respective holders. The use of brand names, product names, common names, trade names, product descriptions etc. even without a particular marking in this work is in no way to be construed to mean that such names may be regarded as unrestricted in respect of trademark and brand protection legislation and could thus be used by anyone.

Coverbild / Cover image: www.ingimage.com

Verlag / Publisher:
Südwestdeutscher Verlag für Hochschulschriften
ist ein Imprint der / is a trademark of
OmniScriptum GmbH & Co. KG
Heinrich-Böcking-Str. 6-8, 66121 Saarbrücken, Deutschland / Germany
Email: info@svh-verlag.de

Herstellung: siehe letzte Seite /
Printed at: see last page
**ISBN: 978-3-8381-5141-0**

Zugl. / Approved by: Duisburg-Essen University,DUE, Diss., 2015

Copyright © 2015 OmniScriptum GmbH & Co. KG
Alle Rechte vorbehalten. / All rights reserved. Saarbrücken 2015

# Abstract

Multiport array-antennas are widely used in beamforming, multiple input multiple output (MIMO), and diversity systems. For applications where a space less than half-wavelength is available as in the case of mobile terminals, strong mutual coupling between antenna ports is introduced. The mutual coupling can cause system performance degradation. A decoupling and matching network could be used to compensate for the effect of mutual coupling. In this thesis, parasitic decoupling elements and eigenmode based decoupling networks are proposed to compensate for the effect of mutual coupling.Both approaches are demonstrated for the monopole four square array antennawhere four monopole elements are placed at the corners of an electrically small platform (chassis) to scan the beam in azimuth.

In the parasitic element decoupling method, parasitic monopole elements are inserted between adjacent and nonadjacent array active elements to compensate for the effect of mutual coupling. On the other hand, the eigenmode based decoupling network is a network connecting the active element ports using reactive components to compensate for the effect of mutual coupling; in practice, the reactive elements are realized in microstrip line.

Recent antenna designs for MIMO favor coupling-element based antenna structures, where the chassis is the main radiator at the lower frequencies. In coupling-element based antenna structures, the coupling element is a nonresonant antenna used to excite the orthogonal characteristic chassis wavemodes through the electro-magnetic field. Making use of the orthogonality of chassis wavemodes, an alternative multiport coupling approach is proposed which decouples the ports. In a design example, a two-port MIMO antenna separately excites the first two dominant chassis wavemodes. Based on the orthogonality of chassis modes, good isolation between antenna ports is achieved.

*To stationed in Al-Aqsa Mosque*

# CONTENTS

**Abstract** .................................................................................................. ii
**Acknowledgment** ................................................................................... iii
**List of Abbreviations** ............................................................................ vii

**1 Introduction** ....................................................................................... 1
    1.1    Background ............................................................................ 1
    1.2    Objectives of the Thesis ......................................................... 2
    1.3    Organization of the Thesis ..................................................... 2

**2 Chassis Wavemodes and Coupling-Element Based Antenna Structures** ........ 5
    2.1    Introduction ............................................................................ 5
    2.2    Theory of Characteristic Modes ............................................. 6
    2.3    Characteristic Modes Radiation Quality Factors .................. 10
    2.4    Numerical Analysis of Characteristic Modes of the Mobile Terminal Chassis ............................................................... 14
    2.5    Coupling Elements ............................................................... 16
    2.6    Design of Impedance Matching Circuits for Coupling-Element Based Antenna Structures ............................................. 19
        2.6.1    Lumped Elements Matching Circuits ....................... 19
        2.6.2    Distributed Elements Matching Circuits .................. 20
    2.7    Broadband Tunable Coupling-Element Based Antenna Structures ........... 22
    2.8    Selective Excitation of Chassis Modes ................................ 25

**3 Study of Mutual Coupling and Chassis Modes Coupling through the Equivalent Circuit Model of Monopoles on a Small Platform** ............... 28
    3.1    Introduction .......................................................................... 28
    3.2    Parallel and Series Resonant Circuits ................................... 29

3.3 Equivalent Circuit Modelling of a Single Monopole on a Small Platform...31
3.4 Equivalent Circuit Modelling of Two Monopoles on a Small Platform......37

**4 Analysis of Mutual Coupling and Chassis Modes Coupling in the Monopole Four Square Array Antenna (MFSAA)..................50**
4.1 Introduction..................50
4.2 Monopole Four Square Array Antenna (MFSAA)..................51
  4.2.1 Radiation Characteristics of the MFSAA..................51
  4.2.2 Directivity of the MFSAA..................53
  4.2.3 Front-to-Back Ratio (F/B) of the MFSAA..................54
  4.2.4 Maximum Absolute Gain of the MFSAA..................54
4.3 Mutual Coupling in the MFSAA..................56
  4.3.1 Mutual Coupling Effect on the MFSAA Radiation Pattern..................59
4.4 Chassis Modes Coupling in the MFSAA..................60
  4.4.1 Chassis Modes Coupling Effect on the MFSAA Radiation Pattern..................63

**5 Compensating for Mutual Coupling and Chassis Modes Coupling in the MFSAA..................67**
5.1 Introduction..................67
5.2 Compensating for Coupled Radiation Patterns of the MFSAA..................68
5.3 Compensating for Port Coupling of the MFSAA..................74
  5.3.1 Parasitic Elements Based Decoupling Technique..................74
  5.3.2 Eigenmode Based Decoupling Network..................82

**6 Utilization of the Characteristic Mode Theory for MIMO Antenna Systems...95**
6.1 Introduction..................95
6.2 Geometry of a Two-Port MIMO Antenna..................96
6.3 Different Two-Port MIMO Antenna Configurations..................99
  6.3.1 Two-Port MIMO Antenna with a Slotted Ground Plane..................99
  6.3.2 Two Port-MIMO Antenna with a Double-Stub Matching Circuit..101
  6.3.3 Two-Port MIMO Antenna with an Offset Feed Point..................104
6.4 Eigenmodal Feed Based Decoupling Network of the MFSAA..................109

| | | |
|---|---|---|
| **7** | **Prototypes: Fabrication and Measurements**..................................................113 | |
| | 7.1 Introduction..................................................................113 | |
| | 7.2 Prototype I: Parasitic Elements Based Decoupling Technique of the MFSAA..................................................................113 | |
| | 7.3 Prototype II: Eigenmode Based Decoupling Network of the MFSAA......116 | |
| | 7.4 Prototype III: Eigenmodal Feed Based Decoupling Network of a Two-Port MIMO Antenna..................................................................122 | |
| **8** | **Conclusion and Future Work**..................................................126 | |
| **References**..................................................................129 | | |
| **Appendices**..................................................................138 | | |

## List of Abbreviations

**MIMO**: Multiple Input Multiple Output
**DOA**: Direction-of-Arrival
**SNR**: Signal to Noise Ratio
**RF**: Radio Frequency
**MFSAA**: Monopole Four Square Array Antenna
**PCB**: Printed Circuit Board
**TCM**: Theory of Characteristic Modes
**MoM**: Method of Moment
**NEC**: Numerical Electromagnetic Code
**CCE**: Capacitive Coupling Element
**ICE**: Inductive Coupling Element
**EM**: Electromagnetic Simulation
**SAR**: Specific Absorption Rate
**PIN**: Positive Intrinsic Negative
**HEMT**: High Electron Mobility Transistor
**CMOS**: Complementary Metal Oxide Semiconductor
**MEMS**: Micro-Electro-Mechanical Switches
**LTCC**: Low-Temperature Co-Fired Ceramic
**ADS**: Advance Design System
**DMN**: Decoupling Matching Network
**EBG**: Electromagnetic Band Gab
**DGS**: Defected Ground Structure
**MSL**: Microstrip Line

# Chapter 1

# Introduction

## 1.1 Background

Smart antennas are divided into two groups, phased array systems (switched beamforming) with a finite number of fixed predefined patterns and adaptive array systems with an infinite number of patterns. In adaptive arrays, the beam is steering in the direction of desired signal while simultaneously nullifying interfering signals, beam direction can be estimated using Direction-of-Arrival (DOA) estimation algorithms. Switched beam system switches between multiple directional fixed beams as the mobile moves throughout the sector. An adaptive array system has better interference rejection capability compared to the switched beam system but has higher complexity and cost.

Multiple Input Multiple Output systems (MIMO) could be considered as an extension to the smart antenna technology. In MIMO, multiple antennas are employed at both ends of the wireless channel (space division multiple access) to significantly enhance the communication system performance. In wireless systems such as mobile communication, the use of multiple antennas can result in a channel capacity and coverage area increase [1]. Traditional communication systems provide the best service under the condition of line-of-sight, while MIMO systems work well under rich scattering condition with multiple independent (uncorrelated) channels.

When two or more antenna elements are close to each other, interchange of energy known as mutual coupling between them occurs. The mutual coupling effect becomes significant as the inter-element spacing decreases. Mutual coupling will result in correlation of channels and significant system performance degradation which causes a reduction in the signal to noise ratio (SNR) of the system [2]. In beamforming antenna arrays, modification of the excitation vector through signal processing can compensate for the effect of mutual coupling on the radiation patterns. However, the SNR of the communication system requires optimization by proper matching of the port impedances. Implementation of RF decoupling networks is required to compensate for SNR degradation. Different decoupling matching networks have been described in the literature to counter for the mutual coupling effect. Eigenmode based decoupling network and parasitic decoupling elements are presented to decouple the monopole four square array antenna (MFSAA). In the MFSAA, four monopole elements are placed in about quarter-wavelength spacing at 2.45 GHz and require equal amplitude excitation with 90° phase shift between array elements to steer the beam in the direction of desired signal and create nulls to suppress interfering signals.

In addition, the effect of characteristic chassis modes is significant in beamforming antenna arrays on small platforms. By the help of the characteristic modes theory, excitation of characteristic current modes and their effect on the array radiation characteristics can be studied. An alternative decoupling network based on the orthogonality of chassis modes is presented. An eigenmodal feed based decoupling network is designed to decouple a two-port MIMO antenna. Two couplers are placed at the middle of the chassis short ends for a selective excitation to the half- and full-wavelength modes of the chassis major axis. A 180° hybrid based feed network was used to feed the structure by the eigenvectors of the selected chassis modes excitation.

## 1.2  Objectives of the Thesis

This thesis deals with design and characterization of antennas on small platforms taking into account the effect of mutual coupling and chassis modes coupling in multi-antenna systems where a space less than half-wavelength is available. Different approaches of decoupling matching networks are presented and compared. Realization of the decoupling networks is developed and verified experimentally.

## 1.3  Organization of the Thesis

This thesis consists of eight chapters, including this introduction chapter. Ch.2 provides background on the characteristic modes theory which is so helpful to analyze the radiation properties of the chassis of a mobile terminal. In particular, it allows understanding of the coupling-element based antenna structures, in which the chassis is considered as the main radiator especially at low frequencies while the antenna works as

## 1.3 Organization of the Thesis

an exciter (coupler) to excite the chassis current modes. Based on the characteristic modes theory, the position of coupling elements on the chassis can be properly chosen for selective excitation of the chassis modes. Single- and multiple-resonances external matching circuits are presented for multiband and broadband coupling-element based antenna structures respectively.

In ch.3, the effect of mutual coupling and chassis modes coupling is studied through the equivalent circuit model of monopoles on a 100 mm × 40 mm chassis. By the help of the equivalent circuit model, calculation of the contribution of chassis modes and coupling element to the total radiation and the effect of chassis modes on the input impedance is presented.

In ch.4, radiation characteristics of the MFSAA are presented. The effect of mutual coupling and chassis modes coupling on the array pattern is analyzed. Simulation results of the current modes distributions and radiation patterns for a 100 mm × 100 mm chassis within the frequency range from 0.5 GHz to 3 GHz are discussed to explain how they affect the MFSAA radiation characteristics.

In ch.5, various decoupling networks are implemented to both correct patterns and decouple ports of the MFSAA. A new excitation vector is derived from the active input impedance of the MFSAA to compensate for coupling effects in the patterns. While eigenmode based decoupling network and parasitic decoupling elements are implemented to decouple the MFSAA ports and thereby enhance the system performance. The effect of chassis modes on the azimuth and elevation element pattern is included and discussed. The excitation vector for the MFSAA with an eigenmode based decoupling network is optimized to allow steering the beam in the direction of desired signal and set the nulls to suppress interfering signals.

In ch.6, an eigenmodal feed based decoupling network is designed based on the orthogonality of the chassis characteristic modes. Different configurations of a two-port MIMO antenna are modelled and compared with respect to the port isolation, frequency band of decoupling, and radiation patterns. Two capacitive coupling elements are placed along the chassis short ends axis to excite the first two dominant modes of a 100 mm × 40 mm chassis. As a second case, the concept of eigenmodal feed based decoupling network is applied to decouple a four-element monopole array (MFSAA). For the MFSAA, the case of an infinite ground plane is considered and the effect of limitation to a 100 mm × 100 mm chassis on the array patterns is analyzed.

In ch.7, verification of the theoretical results is presented. The eigenmode based decoupling network and the parasitic decoupling elements of the MFSAA are fabricated and manufactured, and isolation between ports is measured and compared to simulated results. Two different prototypes of eigenmodal feed based decoupling networks of a two-port MIMO antenna are fabricated. Measured and simulated results for both prototypes are compared.

In ch.8, the most important points are summarized, conclusion for this work is presented and suggestions for future work are given.

# Chapter 2

# Chassis Wavemodes and Coupling-Element Based Antenna Structures

## 2.1 Introduction

Due to the limited space availability on the printed circuit board (PCB) of mobile terminals, a space less than half-wavelength is available between antenna elements. In this case, mutual coupling plays a significant effect on the array performance. In addition, for small platforms, coupling to the chassis modes affects the radiation characteristics of the antenna array. As a demand for multiband small size antennas, the internal self-resonant antennas are replaced by coupling-element based antenna structures, in which the chassis is considered as the main radiator especially at low frequencies while the antenna works as an exciter to excite the chassis wavemodes. Obviously, a good knowledge of the characteristic modes of the chassis is helpful to analyze the effect of chassis modes on the array input impedance and their contribution to the total radiation. In addition, understanding the behavior of coupling element-to-chassis interaction is of interest for selective excitation to chassis modes and how to compensate for the effect of chassis modes on the radiation characteristics of the antenna array. In the next chapter, mutual coupling and chassis modes coupling will be studied through the equivalent circuit model of monopoles on a small platform for deep understanding to the element-to-element and element-to-chassis interaction, then compensating for the effect of mutual coupling and chassis modes coupling could be possible.

## 2.2 Theory of Characteristic Modes

The theory of characteristic modes (TCM) for conducting bodies was developed by Garbacz [3] and an alternative approach was further developed by Harrington [4]. For a conducting body of surface $S$ in an impressed electric field $E^i$, the current density $J$ on $S$ relates to the impressed field via [4]

$$[L(J) - E^i]_{\tan} = 0, \qquad (2.1)$$

where the subscript "tan" denotes the tangential components on $S$. In an antenna problem, the impressed field $E^i$ is the negative of the tangential components of $E$ over $S$, while in a scattering problem, the impressed field $E^i$ is due to a known source external to $S$. In [4], the operator $L$ is defined by

$$L(J) = j\omega A(J) + \nabla \Phi(J), \qquad (2.2)$$

$$A(J) = \mu \oiint_S J(r')\psi(r,r')ds', \qquad (2.3)$$

$$\Phi(J) = \frac{-1}{j\omega\varepsilon} \oiint_S \nabla' \cdot J(r')\psi(r,r')ds', \qquad (2.4)$$

$$\psi(r,r') = \frac{e^{-jk|r-r'|}}{4\pi|r-r'|}, \qquad (2.5)$$

where the field and source points are denoted by $r$ and $r'$ respectively and $\varepsilon, \mu$, and $k$ are the permittivity, permeability, and wavenumber respectively. $A$ and $\Phi$ are the vector potential and the scalar potential respectively. The operator $L$ has the dimension of impedance

$$Z(J) = L(J)_{\tan}. \qquad (2.6)$$

$Z = R + jX$, is a linear symmetric operator, then its Hermitian parts, $R$ and $X$, will be real and symmetric operators given by

$$R = \frac{1}{2}(Z + Z^*), \qquad (2.7)$$

## 2.2 Theory of Characteristic Modes

$$X = \frac{1}{2j}(Z - Z^*). \tag{2.8}$$

The current modes of a conducting body will radiate some power. To get these modes the following eigenvalue problem is considered in [4]

$$Z(J_n) = v_n M(J_n), \tag{2.9}$$

where $v_n$ and $J_n$ are the eigenvalues and eigenmodes respectively. $M$ is a weight operator which has to be chosen. To diagonalize $Z$ and support orthogonal eigenmode patterns, the choice of $M = R$ has to be considered. With $M = R$, $Z = R + jX$ and $v_n = 1 + j\lambda_n$ in (2.9)

$$X(J_n) = \lambda_n R(J_n). \tag{2.10}$$

Real symmetric operators $R$ and $X$ lead to real eigenvalues $\lambda_n$ and eigenmodes $J_n$. The choice $M = R$ gives orthogonality of the eigenmodes which means that $J_n$ should satisfy the orthogonality relationships. Normalizing the amplitudes of the eigenmodes according to $R(J_n)$ gives the orthogonality relations as follows

$$\begin{aligned}
\langle J_m, RJ_n \rangle &= \langle J_m^*, RJ_n \rangle = \delta_{mn}, \\
\langle J_m, XJ_n \rangle &= \langle J_m^*, XJ_n \rangle = \lambda_n \delta_{mn}, \\
\langle J_m, ZJ_n \rangle &= \langle J_m^*, ZJ_n \rangle = v_n \delta_{mn},
\end{aligned} \tag{2.11}$$

where $\delta_{mn}$ is the Kronecker delta given as

$$\delta_{mn} = \begin{cases} 0 & \text{for } m \neq n \\ 1 & \text{for } m = n. \end{cases} \tag{2.12}$$

The eigencurrents or the characteristic currents $J_n$ can be defined as the real currents on the surface $S$ of a conducting body. Eigencurrents $J_n$ and eigenvalues $\lambda_n$ of a conducting body are independent of any specific source or excitation and are computed by reduction of operator equation (2.10) to a matrix equation using the method of moments (MoM) [5] as explained in [6]:

$$[X]J_n = \lambda_n [R]J_n \tag{2.13}$$

The corresponding matrix for the complex eigenvalue is

$$[Z]J_n = (1 + j\lambda_n)[R]J_n, \tag{2.14}$$

where the matrix $[Z]$ is the impedance of the conducting body. The electric field $E_n$ and the magnetic field $H_n$ produced by the characteristic currents $J_n$ on the surface $S$ of a conducting body are called the characteristic fields. The orthogonality properties of the characteristic fields can be obtained from the characteristic currents by means of the complex Poynting theorem, the complex power balance for currents $J$ on $S$ is given by [4]

$$\begin{aligned} P &= \langle J^*, ZJ \rangle = \langle J^*, RJ \rangle + j\langle J^*, XJ \rangle \\ &= \oiint_{S'} E \times H^* \cdot ds + j\omega \iiint_{\tau'} (\mu H \cdot H^* - \varepsilon E \cdot E^*) d\tau \end{aligned} \tag{2.15}$$

where $S'$ is any surface enclosing $S$ and $\tau'$ is the region enclosed by $S'$. Eq. (2.15) is a Hermitian quadratic form, for which the associated Hermitian bilinear form is [4]

$$P(J_m, J_n) = \langle J_m^*, ZJ_n \rangle. \tag{2.16}$$

If $J_m$ and $J_n$ are eigencurrents, then the orthogonality relations (2.11) apply and [4]

$$\oiint_{S'} E_m \times H_n^* \cdot ds + j\omega \iiint_{\tau'} (\mu H_m \cdot H_n^* - \varepsilon E_m \cdot E_n^*) d\tau = (1 + j\lambda_n)\delta_{mn}. \tag{2.17}$$

Due to the orthogonality properties, the characteristic currents $J_n$ can be used as a basis functions to expand the total current $J$ on the surface $S$ of a conducting body. The current $J$ on a conducting surface $S$ is then a linear superposition of the characteristic currents $J_n$ as

$$J = \sum_n a_n J_n \tag{2.18}$$

where $a_n$ are coefficients to be determined. Substituting (2.18) in (2.1) and replacing $L_{\tan}$ by $Z$ will lead to

$$\sum_n a_n Z J_n - E_{\tan}^i = 0 \tag{2.19}$$

## 2.2 Theory of Characteristic Modes

The inner product of (2.19) with $J_m$, where ($m = 1,2,3,...$) gives

$$\sum_n a_n \langle J_m, ZJ_n \rangle - \langle J_m, E^i_{\tan} \rangle = 0 \qquad (2.20)$$

Application of the orthogonality relation (2.11) to (2.20) reduces to

$$a_n v_n = \langle J_n, E^i_{\tan} \rangle, \qquad (2.21)$$

where $v_n = 1 + j\lambda_n$, and the right hand side of (2.21) is called the modal excitation coefficient

$$\langle J_n, E^i_{\tan} \rangle = \oiint_S J_n \cdot E^i_{\tan} ds. \qquad (2.22)$$

Consider the case that the eigencurrents $J_n$ are not normalized and solve for $a_n$ in (2.21)

$$a_n = \frac{\langle J_n, E^i_{\tan} \rangle}{(1 + j\lambda_n)\langle J_n, RJ_n \rangle}. \qquad (2.23)$$

Substituting $a_n$ in (2.18), the modal solution for the current $J$ on the surface $S$ is then

$$J = \sum_n \frac{\langle J_n, E^i_{\tan} \rangle J_n}{(1 + j\lambda_n)\langle J_n, RJ_n \rangle}. \qquad (2.24)$$

The modal excitation coefficient in (2.22) can be considered as an indication of the coupling to the $n$th characteristic mode and determines the contribution of such mode to the total current $J$ on the surface $S$ of a conducting body, while the eigenvalue corresponding to the $n$th mode represents how well the corresponding mode will radiate. As indicated in (2.17), the reactive power is proportional to the magnitude of the eigenvalue. For the case where $|\lambda_n| = 0$, the mode is at resonance because there is no net reactive power. The characteristic modes with positive eigenvalues contribute to store magnetic energy. Therefore those modes are inductive modes, while the characteristic modes with negative eigenvalues contribute to store electric energy and therefore those modes are capacitive modes. Associated with each eigenvalue is a characteristic angle defined as [7]

$$\alpha_n = 180° - \tan^{-1}(\lambda_n). \qquad (2.25)$$

A mode resonates when $|\lambda_n|=0$, meaning that its characteristic angle is $\alpha_n = 180°$. Therefore, when the characteristic angle is close to $180°$, the mode is an effective radiator, while when it's characteristic angle is near $90°$ or $270°$ ($\lambda \to \pm\infty$), the mode is a poor radiator or scatterer, and it mainly stores electric energy for negative $\lambda$ or magnetic energy for positive $\lambda$.

## 2.3 Characteristic Modes Radiation Quality Factors

The radiation quality factor $Q_{rad}$ is the ratio of the time average stored energy to the total radiated energy per cycle [8].

$$Q_{rad} = \frac{\omega W}{P_{rad}}, \qquad (2.26)$$

where $W$ is the time average stored energy in the system, $\omega$ denotes the radian frequency, and $P_{rad}$ is the radiated power. In practice, the $Q_{rad}$ is evaluated at the resonance which implies

$$Q_{rad} = \frac{2\omega_\circ W_{e,m}}{P_{rad}}, \qquad (2.27)$$

where $W_{e,m}$ is either the average stored electric or magnetic energy and $\omega_\circ$ is the resonant frequency. Considering the chassis, which in coupling-element based antenna structures is the main radiator especially at low frequencies, it is of interest to study the contribution of the chassis modes to the total radiation. Since the characteristic modes of the chassis are defined by their corresponding eigenvalues $\lambda_n$ and eigencurrents $J_n$, the radiation quality factor $Q_{rad,n}$ corresponding to the $n$th chassis mode defined by ($\lambda_n, J_n$) will be derived to be proportional to the derivative of the mode eigenvalue $\lambda_n$ with respect to the frequency $\omega$ for the mode at resonance $|\lambda_n|=0$. The following derivation based on [9, 10] will lead to the familiar definition of $Q_{rad,n}$. Starting from the differential form of the complex Poynting theorem for the time harmonic field in an isotropic medium

$$\nabla \cdot \left(\frac{1}{2} E_n \times H_n^*\right) = -\frac{1}{2} J_n^* \cdot E_n - j2\omega\left(\langle w_{m,n}\rangle - \langle w_{e,n}\rangle\right) \qquad (2.28)$$

## 2.3 Characteristic Modes Radiation Quality Factors

where, $w_{m,n}$ and $w_{e,n}$ are the magnetic and electric field energy densities respectively corresponding to the $n$th chassis mode. For an arbitrary antenna, choose $V_0$ such that $\partial V_0$ is coincident with the antenna surface and $V_\infty$ is the region enclosed by a sphere of surface $\partial V_\infty$ lying in the far field region of the antenna. Taking the integration of (2.28) over the region $V_\infty - V_0$ and applying the divergence theorem gives

$$\frac{1}{2}\int_{\partial V_\infty}(E_n \times H_n^*)\cdot ds - \frac{1}{2}V_n I_n^* = -\frac{1}{2}\int_{V_\infty - V_0} \sigma E_n \cdot E_n^* dv - j2\omega \int_{V_\infty - V_0}(\langle w_{m,n}\rangle - \langle w_{e,n}\rangle)dv, \quad (2.29)$$

where $\sigma$ is the conductivity of the medium, and $V_n$ and $I_n$ are the terminal voltage and current corresponding to the $n$th chassis mode. The radiated power $P_{rad,n}$ and the loss power $P_{loss,n}$ are given by

$$P_{rad,n} = \frac{1}{2}\int_{\partial V_\infty}(E_n \times H_n^*)\cdot ds, \quad (2.30)$$

$$P_{loss,n} = \frac{1}{2}\int_{V_\infty - V_0}(\sigma E_n \cdot E_n^*)dv. \quad (2.31)$$

By substituting (2.30) and (2.31) in (2.29), then

$$\frac{1}{2}V_n I_n^* = P_{rad,n} + P_{loss,n} + j2\omega \int_{V_\infty - V_0}(\langle w_{m,n}\rangle - \langle w_{e,n}\rangle)dv. \quad (2.32)$$

Multiplying (2.32) by $(2/|I_n|^2)$ and substituting $(V_n/I_n)$ by the mode complex impedance $Z_n$, this leads to [9]

$$Z_n = R_n + jX_n, \quad (2.33)$$

where

$$R_n = \frac{2P_{rad,n}}{|I_n|^2} + \frac{2P_{loss,n}}{|I_n|^2} \quad (2.34)$$

and

$$X_n = \frac{4\omega(W_{m,n} - W_{e,n})}{|I_n|^2}, \quad (2.35)$$

where $W_{e,n}$ and $W_{m,n}$ are the average stored electric and magnetic field energies corresponding to the $n$ th chassis mode. Define a complex frequency $\tilde{s} = \gamma + j\omega$. In the complex frequency plane, (2.32) can be rewritten as follows [9]

$$\frac{1}{2}\hat{V}_n(\tilde{s})\hat{I}_n^*(\tilde{s}) = \hat{P}_{rad,n} + 2\gamma(\hat{W}_{m,n} + \hat{W}_{e,n}) + 2j\omega(\hat{W}_{m,n} - \hat{W}_{e,n}), \quad (2.36)$$

where the symbol $\hat{\ }$ is used to distinguish the quantities in the complex frequency plane from the corresponding quantities in the frequency domain. For the complex impedance $\hat{Z}_n(\tilde{s})$, multiply (2.36) by ($2/|\hat{I}_n(\tilde{s})|^2$)

$$\hat{Z}_n(\tilde{s}) = \hat{R}_n(\gamma, \omega) + j\hat{X}_n(\gamma, \omega), \quad (2.37)$$

where

$$\hat{R}_n(\gamma, \omega) = \frac{2\hat{P}_{rad,n}}{|\hat{I}_n(\tilde{s})|^2} + \frac{4\gamma(\hat{W}_{m,n} + \hat{W}_{e,n})}{|\hat{I}_n(\tilde{s})|^2} \quad (2.38)$$

and

$$\hat{X}_n(\gamma, \omega) = \frac{4\omega(\hat{W}_{m,n} - \hat{W}_{e,n})}{|\hat{I}_n(\tilde{s})|^2}. \quad (2.39)$$

If $\gamma$ is sufficiently small such that $(\hat{I}_n(\tilde{s}) \approx I_n)$ is independent of $\gamma$. Since the complex impedance is an analytic function its real and imaginary parts satisfy the Cauchy-Riemann conditions

$$\frac{\partial \hat{R}_n(\gamma, \omega)}{\partial \gamma} = \frac{\partial \hat{X}_n(\gamma, \omega)}{\partial \omega}, \quad (2.40)$$

$$\frac{\partial \hat{R}_n(\gamma, \omega)}{\partial \omega} = -\frac{\partial \hat{X}_n(\gamma, \omega)}{\partial \gamma}, \quad (2.41)$$

## 2.3 Characteristic Modes Radiation Quality Factors

from (2.38),

$$\left.\frac{\partial \hat{R}_n(\gamma,\omega)}{\partial \gamma}\right|_{\gamma=0} = \frac{4(W_{m,n}+W_{e,n})}{|I_n|^2} \tag{2.42}$$

from (2.40) and (2.42),

$$\frac{\partial X_n}{\partial \omega} = \left.\frac{\partial \hat{X}_n(\gamma,\omega)}{\partial \omega}\right|_{\gamma=0} = \frac{4(W_{m,n}+W_{e,n})}{|I_n|^2}, \tag{2.43}$$

which is the Foster reactance theorem for antennas. From (2.39), for $(\hat{I}_n(\tilde{s}) \approx I_n)$, and from (2.43), the stored magnetic and electric field energies corresponding to the $n$th chassis mode could be obtained as follows

$$W_{m,n} = \frac{|I_n|^2}{8}\left(\frac{\partial X_n}{\partial \omega} + \frac{X_n}{\omega}\right)$$

$$W_{e,n} = \frac{|I_n|^2}{8}\left(\frac{\partial X_n}{\partial \omega} - \frac{X_n}{\omega}\right). \tag{2.44}$$

By substituting (2.44) in (2.27), the radiation $Q_{rad,n}$ factor will be given by

$$Q_{rad,n} = \frac{|I_n|^2}{4} \frac{\left(\omega \frac{\partial X_n}{\partial \omega} \pm X_n\right)}{P_{rad,n}}$$

$$= \frac{\omega \frac{\partial X_n}{\partial \omega} \pm X_n}{2R_{rad,n}}, \tag{2.45}$$

and from (2.10):

$$Q_{rad,n} = \frac{1}{2}\omega \frac{d\lambda_n}{d\omega} \pm \lambda_n. \tag{2.46}$$

At resonance where ($\omega = \omega_n$), and $\lambda(\omega_n) = 0$,

$$|Q_{rad,n}| = \frac{1}{2}\omega_n \left.\frac{d\lambda_n}{d\omega}\right|_{\omega=\omega_n}. \tag{2.47}$$

## 2.4 Numerical Analysis of Characteristic Modes of the Mobile Terminal Chassis

Numerical evaluation of the characteristic modes of a mobile chassis requires solving the generalized eigenvalue problem in (2.10). In practice, (2.10) need to be reduced into the matrix form (2.13) by applying the MoM as explained in [5]. In [11], the Numerical Electromagnetic Code (NEC2) was extended to solve for the eigenvalue problem in (2.13). The NEC is based on the method of moment solution of the electric field integral equation for thin wires and the magnetic field integral equation for planar surfaces [12]. By the NEC, an arbitrary structure can be approximated by a wire grid structure. In [11], a 100 mm × 40 mm bar type phone chassis was approximated to a wire grid model, Fig.2.1a.

The impedance matrix of the structure $[Z]$ was generated by NEC2 and passed to the LAPACK solver [13] for the real symmetric generalized eigenvalue problem. The calculated frequency dependent eigenvalues in the frequency range from 1 GHz to 3.2 GHz are shown in Fig.2.1b. The first four resonant frequencies of the 100 mm × 40 mm chassis were found as 1.26, 2.68, 2.74 and 3.08 GHz, and their respective radiation quality factors values are 2.3, 3.0, 2.5, and 2.3. The first two modes represent the half- and full-wavelength modes of the chassis major axis, the third mode represents the half-wavelength mode of the chassis minor axis, and the fourth mode represents the half-wavelength mode of the chassis minor and major axis. The current patterns of the first four chassis modes obtained in [11] are shown in Fig.2.2. The blue and red colors correspond to positive and negative signs, while the line thickness indicates the magnitude of the current.

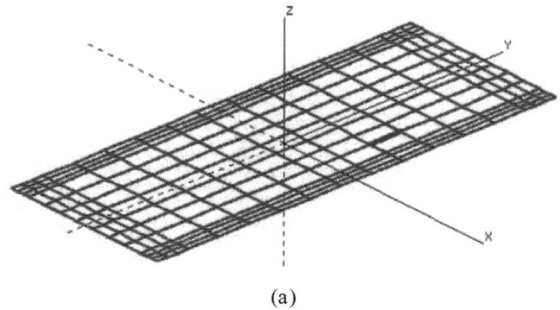

(a)

## 2.4 Numerical Analysis of Characteristic Modes of the Mobile Terminal Chassis 15

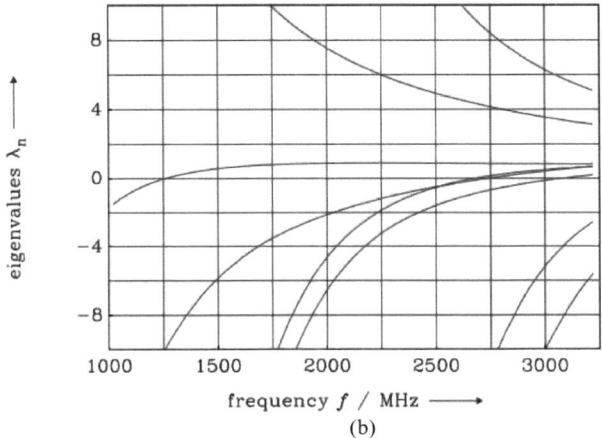

Figure 2.1: (a) The wire grid model of a 100 mm × 40 mm chassis, and (b) the frequency dependent eigenvalues for the first few modes of the chassis [11].

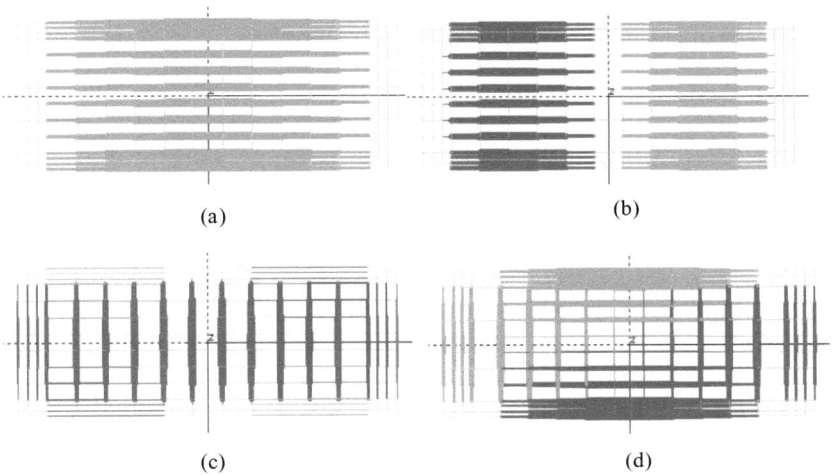

Figure 2.2: The current patterns of the first four 100mm×40mm chassis modes [11].

## 2.5 Coupling Elements

Based on the fact that at low frequencies (around 1 GHz), the chassis works as the main radiator, whereas the antenna element works mainly as a coupling element [14], a simple nonresonant coupling element could be used for efficient excitation of the chassis wavemodes. The concept of coupling-element based antenna makes the size reduction of an internal mobile terminal antenna element possible by replacing a traditional self-resonant antenna element with a simple capacitive or inductive coupling element and an external matching circuit. Coupling-element based antennas consist mainly of three parts: The chassis of the mobile terminal which works as the main radiator of the antenna-chassis combination especially at low frequencies, the coupling element with proper shape and location to support maximum coupling to the chassis characteristic wavemodes, and an external matching circuit to 50 $\Omega$ can be considered as the third part of the combined structure.

According to [14], the relative bandwidth of the combination of antenna element and chassis depends on the coupling between the antenna element and the chassis: As the coupling to the chassis wavemodes increases, the relative bandwidth also increases.

For optimal coupling to the chassis modes, the coupling element location and shape should be correctly chosen. The coupling element can be either a capacitive coupling element (CCE) that couples to the electric field of the chassis or an inductive coupling element (ICE) that couples to the magnetic field of the chassis.

In [15], the effect of the capacitive coupling element location on a $100\,\text{mm} \times 40\,\text{mm}$ chassis was studied by using electromagnetic simulations. A probe of length 3 mm was placed at different locations on a $100\,\text{mm} \times 40\,\text{mm}$ chassis. The radiation quality factor $Q_{rad}$ was obtained from the EM simulation of the input admittance as the ratio of susceptance to conductance.

The minimum radiation quality factors were obtained at the corner and short ends of the chassis. The minimum radiation quality factor represents the strongest coupling to the chassis wavemodes. Fig.2.3 shows the normalized simulated radiation quality factor at 920 MHz obtained with a small probe placed at different locations on a chassis of $100\,\text{mm} \times 40\,\text{mm}$ [15].

The behavior of the radiation quality factor in Fig. 2.3 can be explained by the half-wave dipole type current distribution of a typical mobile handset chassis at 900 MHz [14]. The electric field minimum is located at the center of the chassis, while the electric

## 2.5 Coupling Elements

field maximums are located at the chassis short ends. The quality factor curves in Fig.2.3 vary inversely proportional to the electric field on the chassis.

In [16, 17], the theory of characteristic modes has been applied for optimal coupling element location on a mobile terminal chassis. The bandwidth potential (6 dB bandwidth) of a small center-fed patch antenna as a function of frequency for different locations at the corner, at the center, and at the midpoints of the long and short ends of a 100 mm × 40 mm chassis has been presented in [16]. The maximum bandwidth potential was obtained when the coupling element is placed at the corner of the chassis, while minimum bandwidth potential was obtained when the coupling element is placed at the center of the chassis where the electric field is minimum[15].

On the other hand, the coupling element shape has to be properly chosen for strong coupling to the chassis dominant wavemodes. In [14, 18], the coupling element is bent over the shorter end of a 100 mm × 40 mm chassis for stronger capacitive coupling to the half-wavelength (odd mode) and full-wavelength (even mode) of the chassis major axis, Fig.2.4a. In [19], another two different configurations for capacitive coupling element were presented. In the first configuration the CCE is bent around a 100 mm × 40 mm chassis short end to enhance coupling to the odd and even chassis modes of the major axis, Fig.2.4b.

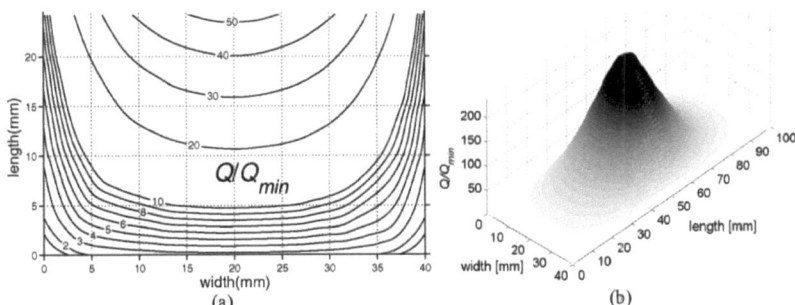

Figure 2.3: Normalized ($Q_{min} = 195$) simulated radiation quality factors obtained with a small probe moved on a 100 mm × 40 mm chassis. (a) Closer view of the shorter end of the chassis and (b) 3-D view of the whole chassis [15].

While in the second configuration, the CCE is bent around the chassis corner to strongly couple to the dominant four characteristic modes of the chassis, Fig.2.4c. Furthermore, inductive coupling elements to couple to the chassis modes through the magnetic field instead of the electric field have been presented in [20, 21], where a thin wire loop is used as an inductive coupling element to couple to the chassis current modes through the magnetic field, Fig.2.4d.

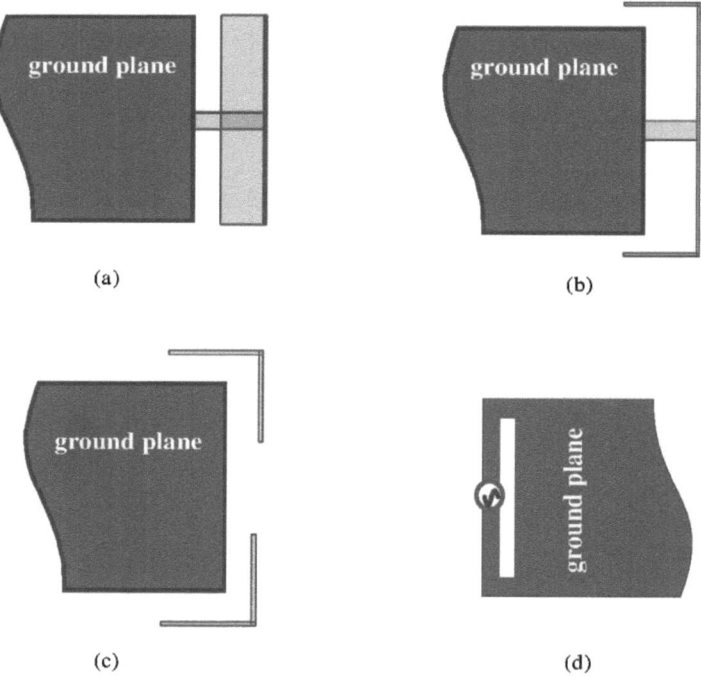

Figure 2.4: Different configurations of capacitive and inductive coupling elements.

## 2.6 Design of Impedance Matching Circuits for Coupling-Element Based Antenna Structures

As mentioned before a coupling element is a nonresonant element, which can be matched at any frequency by using an external matching circuit [22]. Design of a matching circuit to match a complex load to a desired resonance can be implemented in two ways, using lumped reactive elements or distributed elements.

### 2.6.1 Lumped Elements Matching Circuits

A complex load $Z_L = R_L + jX_L$ can be matched to $Z_\circ = 50\,\Omega$ by two lumped elements (L-section network) with two different configurations based on the load impedance value compared to $Z_\circ = 50\,\Omega$ [23]. The two different configurations of L-section (single resonance) matching networks are shown in Fig.2.5. The reactance $X$ and susceptance $B$ of Fig.2.5a configuration are given as [23]

$$X = \pm\sqrt{R_L(Z_\circ - R_L)} - X_L \qquad (2.48)$$

$$B = \pm\frac{\sqrt{(Z_\circ - R_L)/R_L}}{Z_\circ}. \qquad (2.49)$$

Respectively, the reactance $X$ and susceptance $B$ of Fig.2.5b configuration are given as [23]

$$B = \frac{X_L \pm \sqrt{R_L/Z_\circ}\sqrt{R_L^2 + X_L^2 - Z_\circ R_L}}{R_L^2 + X_L^2} \qquad (2.50)$$

$$X = \frac{1}{B} + \frac{X_L Z_\circ}{R_L} - \frac{Z_\circ}{BR_L}. \qquad (2.51)$$

The lumped elements may be inductors or capacitors. Positive reactance $X$ or negative susceptance $B$ means inductance, while positive susceptance $B$ or negative reactance $X$ means capacitance. Thus there are eight possibilities of the matching circuit for various load impedances.

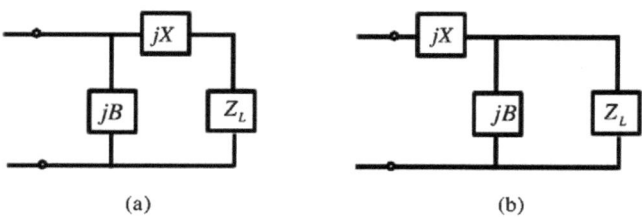

(a)     (b)

Figure 2.5: L-section matching circuits. (a) Matching circuit for $R_L < 50\,\Omega$ and (b) matching circuit for $R_L > 50\,\Omega$.

### 2.6.2 Distributed Elements Matching Circuits

To match a complex load $Z_L = R_L + jX_L$ to a transmission line of characteristic impedance $Z_o = 50\,\Omega$, a subsection of a transmission line of length $l_s$ and characteristic impedance $Z_o$ has to be placed in series (series stub) or in parallel (shunt stub) at a distance $d$ from the load $Z_L$. Fig.2.6 shows a single shunt stub used to match the load impedance $Z_L$ to the transmission line characteristic impedance $Z_o$. The distance $d$ in Fig.2.6 for $R_L \ne Z_o$ can be obtained from [23]

$$\tan\beta d = \frac{X_L \pm \sqrt{R_L[(Z_o - R_L)^2 + X_L^2]/Z_o}}{R_L - Z_o} \quad \text{for } R_L \ne Z_o, \qquad (2.52)$$

where $\beta$ is the propagation constant for a lossless transmission line. For $R_L = Z_o$, $\tan\beta d = -X_L/(2Z_o)$. The required susceptance $B_s$ of the stub is [23]

$$B_s = \frac{(Z_o - X_L t)(X_L + Z_o t) - R_L^2 t}{Z_o[R_L^2 + (X_L + Z_o t)^2]}, \qquad (2.53)$$

where $t = \tan\beta d$. The length of a short-circuited stub $l_s/\lambda$ is given by

$$\frac{l_s}{\lambda} = \frac{-1}{2\pi}\tan^{-1}\left(\frac{Y_o}{B_s}\right), \qquad (2.54)$$

while for an open-circuited stub,

## 2.6 Design of Impedance Matching Circuits for CE Based Antenna Structures

$$\frac{l_s}{\lambda} = \frac{1}{2\pi} \tan^{-1}\left(\frac{B_s}{Y_o}\right), \quad (2.55)$$

where the characteristic admittance of the transmission line $Y_o = 1/Z_o$.

For real load impedances, a quarter-wave transformer matching circuit will be a good matching technique. Fig.2.7 shows a quarter-wave transformer with characteristic impedance $Z_1$ given by [23]

$$Z_1 = \sqrt{Z_o R_L} \quad (2.56)$$

For complex load impedance, it can be transformed to a real one by using a tunable length of a transmission line between the load and the quarter-wave transformer.

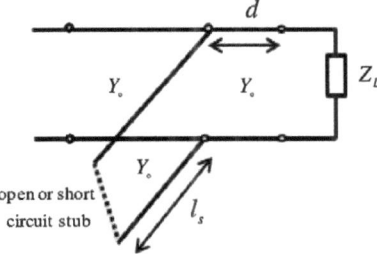

Figure 2.6: A single shunt stub matching circuit of length $l_s$ placed at distance $d$ from the load.

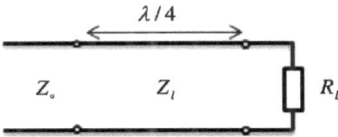

Figure 2.7: A quarter-wave matching transformer.

## 2.7 Broadband Tunable Coupling-Element Based Antenna Structures

The use of multiple-resonances matching circuits is an effective way to enhance the impedance bandwidth of a coupling-element based antenna structure. For matching with distributed elements, double-stub or multiple sections of quarter-wave transformer will be effective for the purpose of multiple-resonances or wider bandwidth. While for matching with lumped elements, multiple L-section network can be used for multiple-resonances or bandwidth enhancement. Two stages of L-section network can be modeled as a $\pi$- or $T$-matching circuit. Multi-stages matching circuits provide extra degree of freedom to control the bandwidth for a wide scope of tuning.

A wider bandwidth could also be obtained by modifying the ground plane, such that no ground plane metallization remains under the coupling element itself (off-ground coupling element) [24, 25]. The drawback of an off-ground coupling element is higher absorption rate (SAR) -if beside the head- in comparison to an on-ground coupling element. In the talking position, the SAR was calculated at 900 MHz for a coupling element placed beside the user ear. A SAR of 1.46 W/Kg was obtained for the case of an on-ground coupling element, while a SAR of 1.92 W/Kg was obtained for the case of an off-ground coupling element [25].

The nonresonant coupling-element based antennas are good candidates from two points of view: Allowance of simultaneous impedance matching at multiple frequency bands (broadband matching) [26, 27] and frequency reconfigurable multi-band antennas (switchable narrowband matching) [28-31]. To implement multiple frequency bands, two different arrangements of coupling element(s) and matching circuits are considered. In the first approach, each desired frequency band could be obtained by a coupling element with a separate matching circuit [26], while in the second approach one coupling element with multiple feed points is used [27] where each feed point is connecting to a matching circuit corresponding to the desired frequency band. The two different approaches are shown in Fig.2.8a and 2.8b.

On the other hand, for frequency reconfigurable coupling-element based antenna, one coupler with multiple matching circuits is required and an RF switch between the matching circuits and the coupler is needed to switch between the multiple-frequency bands to support reconfigurability.

## 2.7 Broadband Tunable Coupling-Element Based Antenna Structures

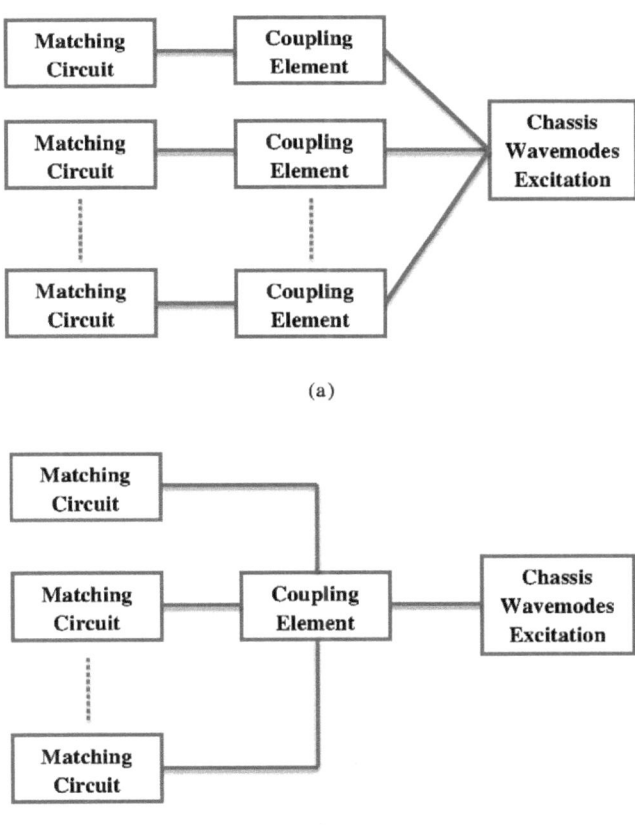

Figure 2.8: Different approaches of multiband coupling-element based antennas. (a) Several coupling elements are connected to several matching circuits. (b) Several matching circuits are connected to a single coupling element with multi-feed points.

For more clarity, a coupling element with multiple matching circuits modelling the frequency reconfigurable approach is presented in Fig.2.9. The low input resistance of the coupling element makes the resistance across the switch next to the coupling element relatively high resulting in a high power loss. In [30], a shunt inductor between the coupling element and the switch improves the total efficiency of the system.

RF switches could be semiconductor switches (e.g. PIN diode switches, HEMT switches and CMOS switches) or micro-electromechanical switches (MEMS). The MEMS switches compared to the semiconductor switches have very low insertion loss, very low power consumption, very high isolation, and very wide operating frequency range [32, 33]. The drawback of the MEMS switches is the high cost compared to the very low cost of the semiconductor switches, and the life cycle reliability of the MEMS switches is still insufficient for the use in mobile handsets.

The advantage of the switchable narrowband matching over the simultaneous multiband matching is the simplicity of the matching network. In simultaneous multiband matching, by increasing the number of lumped elements in the matching circuit, wider bandwidth is obtained, while on the other hand the complexity and RF losses of the circuit will be increased. Recently, high-Q inductors, capacitors, transmission lines, and other passive elements can be integrated into multilayer low-temperature co-fired ceramic (LTCC) substrates occupying a very small volume [34, 35].

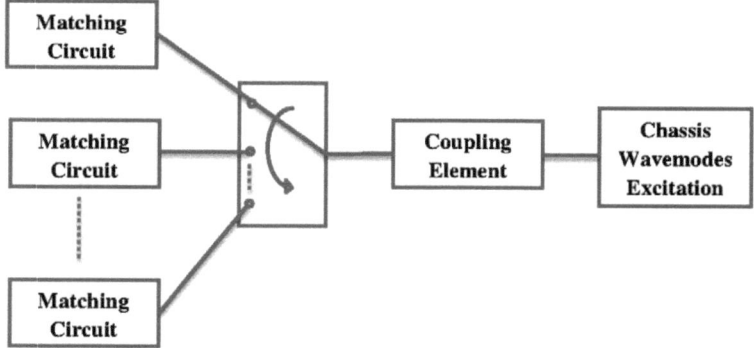

Figure 2.9: Frequency reconfigurable multiband antenna.

## 2.8 Selective Excitation of Chassis Modes

For selective excitation to the chassis modes, the coupling element shape and location on the chassis have to be properly chosen. The optimal coupling element location after [14, 16, 17] is at the positions where the electric field of the chassis has its maximum. In [11], a numerical analysis for a chassis of dimensions 100 mm × 40 mm was presented and the current distributions for the dominant chassis modes were obtained. According to [11, 14, 16, 17], selective excitation to the half- and full-wavelength modes of the chassis major axis required a coupling element to be placed at the middle point of the chassis short end, where the electric field for the half- and full-wavelength modes of the chassis major axis is at its maximum, while it is at its minimum for the half-wavelength mode of the chassis minor axis.

As an example, a CCE is properly designed and placed at the midpoint of the chassis short end for efficient selective excitation to the half- and full-wavelength modes of the chassis major axis. The chassis dimensions are 100 mm × 40 mm. The whole structure is shown in Fig.2.10. The coupling element is bent around the chassis short end to strongly couple to the chassis modes. The CCE is an off-ground element for the purpose of bandwidth enhancement [24, 25] with a ground clearance of 5mm. The dimensions of the whole structure are $105 \times 48 \times 5 mm^3$ , Fig.2.10.

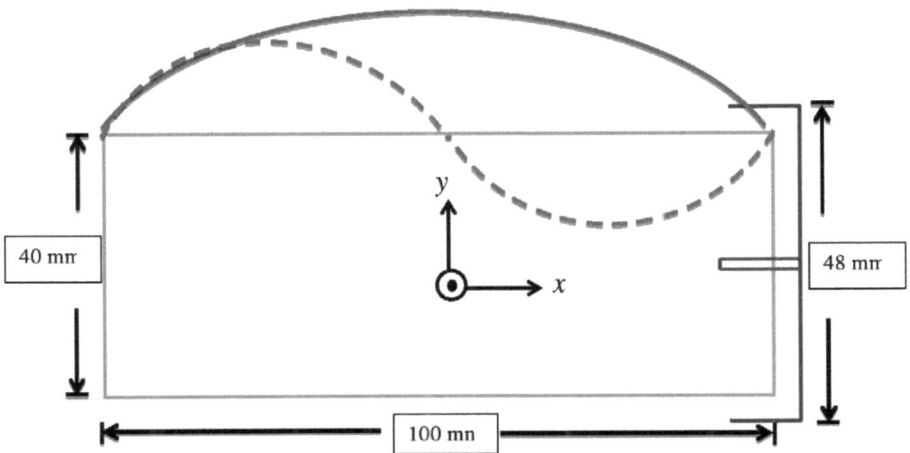

Figure 2.10: Capacitive coupling-element based antenna structure.

The CCE element is matched by an external single resonant L-section matching circuit at each simulated frequency point between 0.5 GHz and 3 GHz. The relative bandwidth (10 dB bandwidth) obtained at each simulated frequency point is plotted in Fig.2.11. The half- and full-wavelength modes of the chassis major axis can clearly be seen at 0.9 GHz and 2.2 GHz respectively. The contribution of the chassis is significant at around 1 GHz, while it is lower at around 2 GHz, as expected [14].

Using multiple-resonances matching circuit can enhance the bandwidth of the coupling-element based antenna structure. A multiple resonances external matching circuit is designed for the structure in Fig.2.10. The relative bandwidth (10 dB bandwidth) as a function of the lumped elements number in the matching circuit is plotted in Fig.2.12 at the resonant frequencies of the half- and full-wavelength modes of the chassis major axis. According to Fig.2.12, the relative bandwidth increases with the number of lumped elements in the external matching circuit.

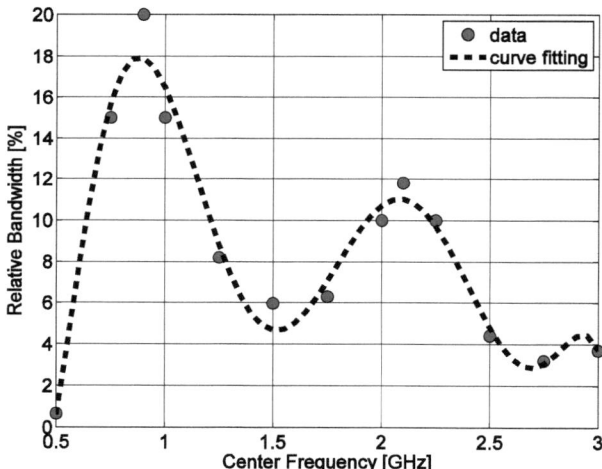

Figure 2.11: Relative bandwidth (10 dB bandwidth) calculated when the CCE in Fig.2.10 is matched by an external optimal L-section matching circuit.

## 2.8 Selective Excitation of Chassis Modes

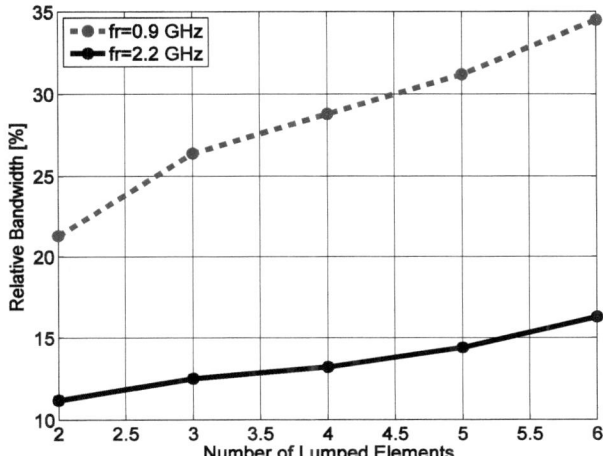

Figure 2.12: Relative bandwidth (10 dB bandwidth) calculated vs. the number of lumped elements in the matching circuit of the CCE in Fig.2.10.

# Chapter 3

# Study of Mutual Coupling and Chassis Modes Coupling through the Equivalent Circuit Model of Monopoles on a Small Platform

## 3.1 Introduction

In coupling-element based antenna structures, the coupling element is used as an exciter to excite the chassis modes. An overall understanding of how different parts of the combined structure contribute to the input impedance and total radiation could be possible through a lumped elements equivalent circuit model based on parallel and series coupled resonators to model the chassis wavemodes and the antenna elements wavemodes. The work in [14, 36-38] provides the basis to model the antenna-chassis combination as a parallel and series coupled resonators. In [39, 40], an equivalent circuit of antenna-chassis combination with a single coupling element was modeled. While in [41, 42], an equivalent circuit model of antenna-chassis combination with two coupling elements was designed and analyzed. The resonant frequencies and the radiation quality factors of the coupled resonators in the equivalent circuit model could be obtained from the theory of characteristic modes. In [11], by the help of characteristic mode theory, a 100 mm × 40 mm bar type phone chassis was approximated to a wire grid model and the first four resonant frequencies of the chassis and their respective radiation quality factors values were numerically calculated. In order to model properly the mutual coupling in the equivalent circuit model in case of antenna-chassis combination of two or more coupling elements, the work in [43] will be a good start point.

## 3.2 Parallel and Series Resonant Circuits

A parallel RLC resonant circuit is shown in Fig.3.1a, where the input impedance $Z_{in}$ is given by [23]

$$Z_{in} = \left( \frac{1}{R} + \frac{1}{j\omega L} + j\omega C \right)^{-1}, \qquad (3.1)$$

and the complex power delivered to the resonator is

$$P_{in} = \frac{1}{2}|V|^2 \left( \frac{1}{R} + \frac{j}{\omega L} - j\omega C \right). \qquad (3.2)$$

The average electric and magnetic stored energies $W_e$ and $W_m$ respectively are given by

$$W_e = \frac{1}{4}|V|^2 C \qquad (3.3)$$

$$W_m = \frac{1}{4}|V|^2 \frac{1}{\omega^2 L}. \qquad (3.4)$$

While the power dissipated by $R$ is

$$P_{loss} = \frac{1}{2}\frac{|V|^2}{R}. \qquad (3.5)$$

The input impedance $Z_{in}$ can be rewritten as

$$Z_{in} = \frac{P_{loss} + 2j\omega(W_m - W_e)}{\frac{1}{2}|I|^2}. \qquad (3.6)$$

Resonance occurs when the input impedance $Z_{in}$ is purely real ($W_m = W_e$). The resonant frequency $\omega_\circ$ is then defined as

$$\omega_\circ = \frac{1}{\sqrt{LC}}. \qquad (3.7)$$

The radiation quality factor $Q_{rad}$ for a parallel resonator is defined as the ratio of average stored energy in the resonator and energy dissipated per cycle. At resonance $Q_{rad}$ is given by

$$Q_{rad} = 2\omega_o \frac{W_{e,m}}{P_{loss}}$$
$$= \omega_o RC, \qquad (3.8)$$

where $W_{e,m}$ is either the average electric or magnetic stored energy.

A series RLC resonant circuit is shown in Fig.3.1b, the input impedance $Z_{in}$ is

$$Z_{in} = \left( R + j\omega L + \frac{1}{j\omega C} \right), \qquad (3.9)$$

and the complex power delivered to the resonator is

$$P_{in} = \frac{1}{2} Z_{in} |I|^2. \qquad (3.10)$$

The average electric and magnetic stored energies $W_e$ and $W_m$ respectively are given by

$$W_e = \frac{1}{4}|I|^2 \frac{1}{\omega^2 C} \qquad (3.11)$$

$$W_m = \frac{1}{4}|I|^2 L. \qquad (3.12)$$

Resonance occurs when the average stored electric and magnetic energies are equal ( $W_m = W_e$ ). The resonant frequency $\omega_o$ and the radiation quality factor $Q_{rad}$ at resonance for a series resonator are given by

$$\omega_o = \frac{1}{\sqrt{LC}} \qquad (3.13)$$

$$Q_{rad} = \frac{1}{\omega_o RC}. \qquad (3.14)$$

## 3.3 Equivalent Circuit Modelling of a Single Monopole on a Small Platform

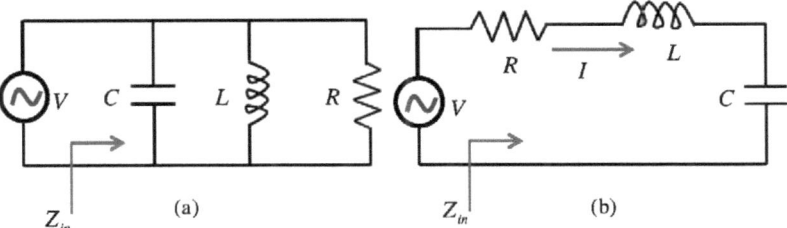

Figure 3.1: Circuit model for (a) parallel RLC resonator and (b) series RLC resonator.

### 3.3 Equivalent Circuit Modelling of a Single Monopole on a Small Platform

For a 100 mm × 40 mm chassis, the first four resonant frequencies of the chassis were numerically calculated as 1.26, 2.68, 2.74 and 3.08 GHz, and their respective radiation quality factors values are 2.3, 3.0, 2.5, and 2.3 [11]. The first two modes represent the half- and full-wavelength modes of the chassis major axis, the third mode represents the half-wavelength mode of the chassis minor axis, and the fourth mode represents the half-wavelength mode of the chassis minor and major axis. A single quarter-wavelength monopole at 2.45 GHz is used to excite the chassis modes.

For optimal excitation of the chassis modes, the exciter location has to be properly chosen. In [16, 17], the maximum bandwidth potential was obtained when the coupling element is placed at the corner of the chassis, while minimum bandwidth potential was obtained when the coupling element is placed at the center of the chassis where the electric field is minimum [15].

In the equivalent circuit model, the general case in which the monopole is placed at the corner of the chassis (100 mm × 40 mm) will be considered. Based on [37], a series RLC resonator is used to model the monopole mode. Consequently, a parallel RLC resonator is used to model the chassis modes [14]. The resonant frequencies and corresponding quality factors calculated in [11] are used in the design of the equivalent circuit as initial values. Due to the electrical length increase of the chassis, the monopole (exciter) slightly tunes downward the resonant frequencies of the chassis modes [44]. For a frequency range between 0.5 GHz and 3GHz, the first three chassis modes will be considered. The equivalent circuit model then includes a series RLC resonator to model the monopole wavemode, and three cascaded parallel RLC resonators to model the chassis modes, Fig.3.2. The coupling between the monopole and the chassis modes is modeled as an ideal transformer [39].

Optimization goals in the equivalent circuit model include conditions on the resonant frequencies and $Q$-factors of both monopole and chassis modes as well as good agreement between EM simulation and circuit simulation using the Advanced Design System (ADS) simulator. The equivalent circuit model in ADC of a single monopole on a $100\,\text{mm} \times 40\,\text{mm}$ chassis is shown in Appendix A.

The monopole radiation resistance is modeled as a frequency dependent resistance for better fitting to the monopole impedance behavior. The parasitic capacitor $C_p$ in the equivalent circuit model is optimized for good agreement with EM simulation. As can be seen from Fig.3.3, the equivalent circuit model results are in a good agreement with the EM simulation results. The 1$^{st}$ chassis mode is clearly seen as a peak in the monopole input resistance at the 1 GHz frequency range. The effect of the 2$^{nd}$ and 3$^{rd}$ chassis modes is not clearly seen due to the dominance of the quarter-wavelength monopole resonance around 2.5 GHz.

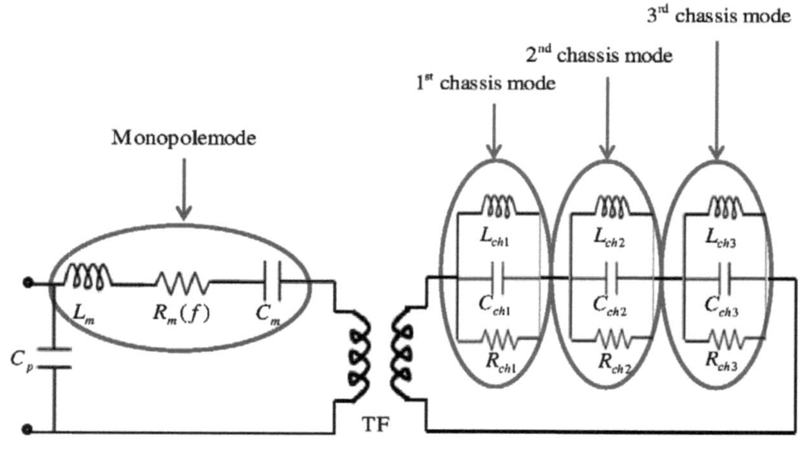

Figure 3.2: Equivalent circuit model of a single monopole placed at the corner of a $100\,\text{mm} \times 40\,\text{mm}$ chassis.

## 3.3 Equivalent Circuit Modelling of a Single Monopole on a Small Platform 33

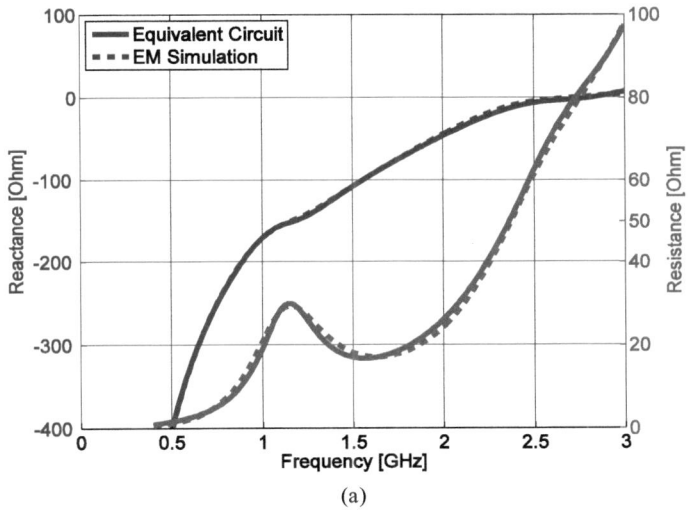

(a)

Reflection coefficient S11

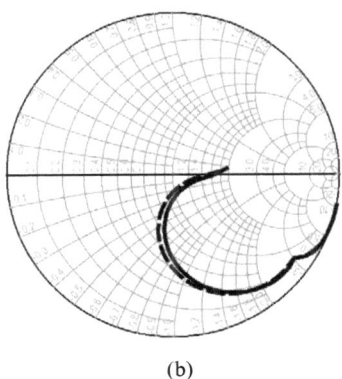

(b)

Figure 3.3: (a) Input impedance and (b) reflection coefficient S11 in Smith chart for a single monopole placed at the chassis corner from equivalent circuit model (solid) and EM simulation (dash).

By the help of the equivalent circuit model, the contribution of each part of the antenna-chassis combination to the input impedance and total radiation can be evaluated. In Fig.3.4, the effect of chassis modes in the monopole input impedance can be seen. The behavior of the monopole input resistance is strongly influenced by the chassis modes, while the reactance behavior is mostly determined by the monopole mode. The $1^{st}$ chassis mode is clearly seen as a peak at the 1 GHz frequency range, while the $2^{nd}$ and the $3^{rd}$ chassis modes increase the resistance at the 2.5 GHz frequency range where the quarter-wavelength monopole is at resonance.

The percentage contribution of each chassis mode and monopole mode to the total radiation can also be calculated by the equivalent circuit model as shown in Fig.3.5. The percentage contribution of each chassis mode is calculated as the ratio of the chassis mode power and the total power of the antenna-chassis combination. It can be seen that at the 1 GHz frequency range, 85% of the power is radiated by the $1^{st}$ chassis mode, while at the 2.5 GHz frequency range, around 35% of the power is radiated by the $2^{nd}$ and the $3^{rd}$ chassis modes. At 2 GHz, the quarter wavelength monopole radiates 70% of the total power.

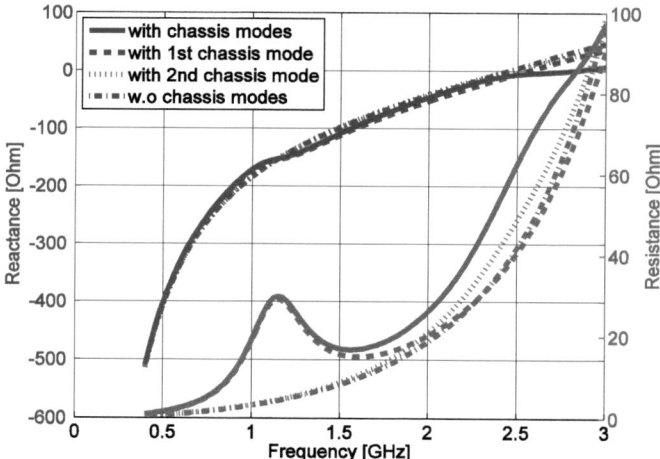

Figure 3.4: Effect of chassis modes on the input impedance of a single monopole placed at the corner of a 100 mm × 40 mm chassis.

## 3.3 Equivalent Circuit Modelling of a Single Monopole on a Small Platform

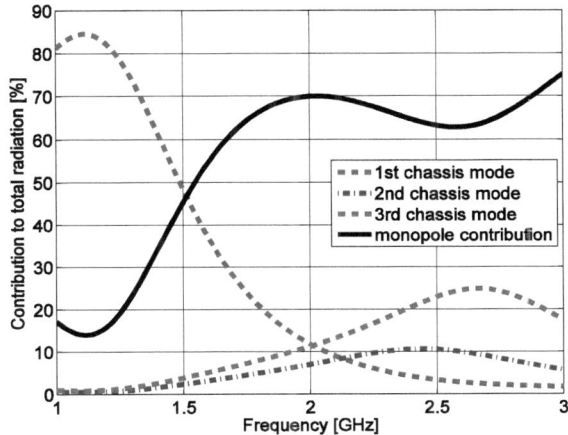

Figure 3.5: Contribution of chassis modes to the total radiated power for a single monopole placed at the corner of a 100 mm × 40 mm chassis

For a selective excitation of chassis modes, the monopole is placed at different positions on a 100 mm × 40 mm chassis: At the corner where the first three chassis modes are excited, at the middle of the chassis short end where the first two chassis modes are excited, and at the center of the chassis where only the $2^{nd}$ chassis mode is excited. The results from the equivalent circuit for different positions of a single monopole on a 100 mm × 40 mm chassis are in a good agreement with that from EM simulation as can be seen from Fig.3.6a and Fig.3.6b.

The values for the RLC resonators, the resonant frequencies and the radiation quality factors resulting from the optimization process in ADS for different positions of a single monopole on a 100 mm × 40 mm chassis are summarized in Tab. 3.1. The value of the resistance in the parallel RLC resonators modelling the individual chassis mode corresponds to the excitation of the modes. For the case of a single monopole placed at the center of the chassis, the first and the third chassis modes are not excited [16, 17]. The same result could be obtained from the equivalent circuit model, low resistance and high capacitance show that the corresponding chassis mode is poorly excited and is more probable to store energy rather than to radiate.

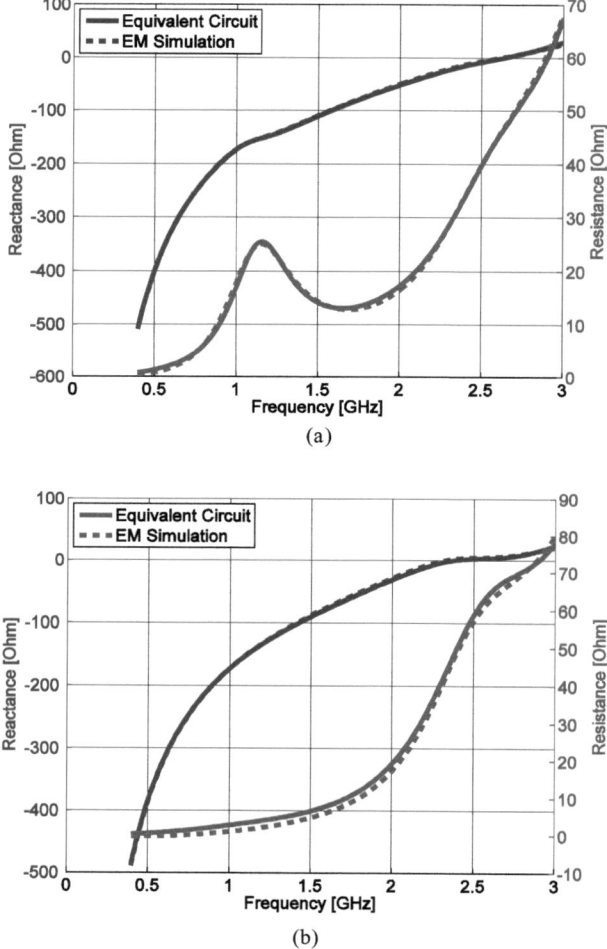

Figure 3.6: Input impedance for a single monopole placed (a) at the middle of the chassis short end, and (b) at the center of the chassis from the equivalent circuit model (solid) and EM simulation (dash).

TABLE 3.1: RESONATOR PROPERTIES FOR LOADED AND UNLOADED CHASSIS.

| Chassis modes | | Loaded chassis | | | Unloaded chassis | |
|---|---|---|---|---|---|---|
| | | corner | center | middle-short end | $f_r$ GHz | $Q$ |
| 1st chassis mode | $C$ pF | 7.956 | 55.028 | 7.527 | 1.26 | 2.3 |
| | $L$ nH | 2.398 | 0.354 | 2.507 | | |
| | $R$ Ω | 48.561 | 2.538 | 45.448 | | |
| | $f_r$ GHz | 1.15 | 1.14 | 1.16 | | |
| | $Q$ | 2.79 | 1 | 2.49 | | |
| 2nd chassis mode | $C$ pF | 23.164 | 4.577 | 16.595 | 2.68 | 3.0 |
| | $L$ nH | 0.164 | 0.806 | 0.226 | | |
| | $R$ Ω | 6.676 | 33.696 | 11.848 | | |
| | $f_r$ GHz | 2.58 | 2.62 | 2.60 | | |
| | $Q$ | 2.51 | 2.54 | 3.2 | | |
| 3rd chassis mode | $C$ pF | 9.756 | 64.671 | 63.73 | 2.74 | 2.5 |
| | $L$ nH | 0.351 | 0.053 | 0.054 | | |
| | $R$ Ω | 17.656 | 2.495 | 2 | | |
| | $f_r$ GHz | 2.72 | 2.72 | 2.71 | | |
| | $Q$ | 2.94 | 2.76 | 2.17 | | |

## 3.4 Equivalent Circuit Modelling of Two Monopoles on a Small Platform

In comparison with the equivalent circuit of a single monopole on a small platform, in addition to model and analyze the coupling between the monopole elements and the chassis, the mutual coupling between the two monopoles has to be modelled properly. The geometry of two quarter-wavelength monopoles at 2.45 GHz mounted on a 100 mm × 40 mm ground plane and placed at distance $d$ from the middle of the chassis short ends is shown in Fig.3.7. For selective excitation of the chassis modes, positions and excitation phases of the two monopole coupling elements have to be properly chosen. In Fig.3.7, to excite the half- and full-wavelength modes of the chassis major axis, odd and even excitation, respectively, of the two monopoles placed at the middle of the chassis short ends $d = 4$ mm is required. The unitary matrix for such a structure is given by

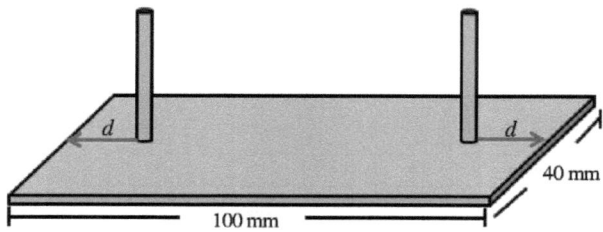

Figure 3.7: Two quarter-wavelength monopoles mounted on a 100 mm × 40 mm ground plane.

$$U = \frac{1}{\sqrt{2}} \begin{bmatrix} 1 & 1 \\ 1 & -1 \end{bmatrix} \quad (3.15)$$

The two-column vectors of U represent the excitation vectors for the half- and full-wavelength chassis modes. In-phase excitation is needed to excite the $2^{nd}$ chassis mode (even excitation), while anti-phase excitation is needed to excite the $1^{st}$ chassis mode (odd excitation). The surface current distributions from EM simulation using Empire Xccel Fig.3.8 of the odd and even excitation of the two monopoles placed at the middle of the chassis short ends with $d = 4$ mm exhibit close similarity with the odd and even characteristic current modes of the chassis.

The effect of chassis modes coupling on the monopole input impedance is shown in Fig.3.9. The reactance behavior of the impedance below the quarter-wave resonance is mainly determined by the monopole self-impedance, while the effect of the chassis modes is dominant in the resistance part of the impedance. This can be explained by the high Q-factor of the monopole compared to the chassis modes and the low radiation resistance of the monopole.

## 3.4 Equivalent Circuit Modelling of Two Monopoles on a Small Platform

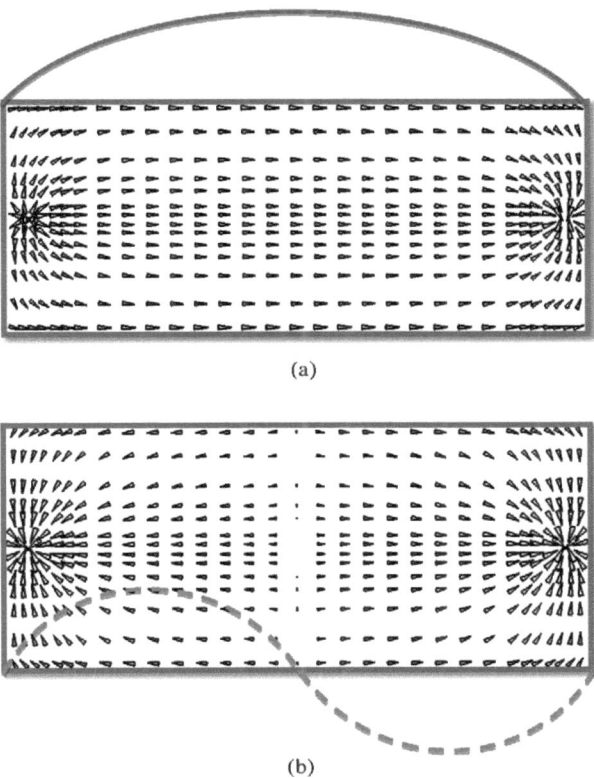

Figure 3.8: Current distribution for two monopoles at the middle of a 100 mm × 40 mm chassis short ends. (a) Odd excitation at 1.14 GHz and (b) even excitation at 2.45 GHz.( Arrows indicate the current direction)

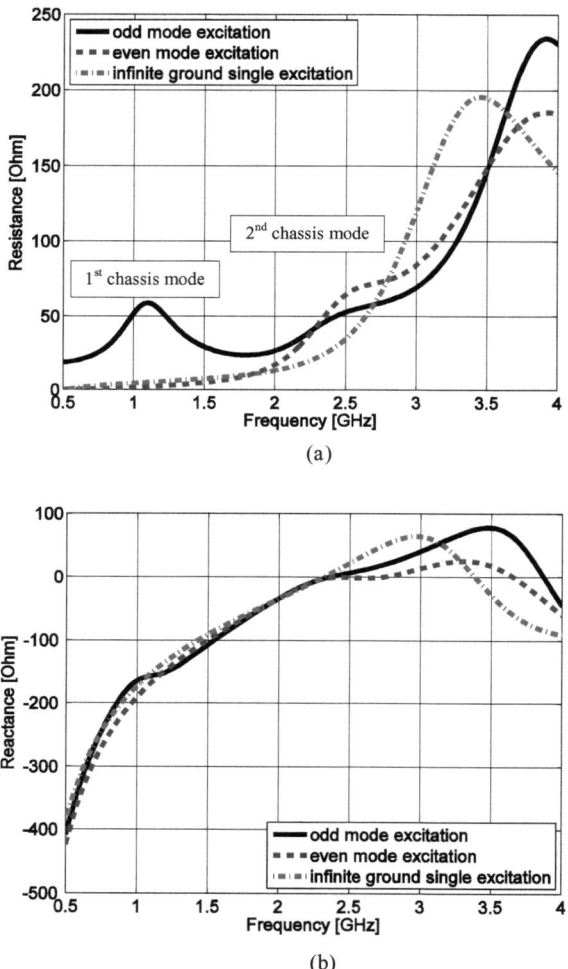

Figure 3.9: Active input impedance of the simulated monopoles at $d = 4$ mm for odd and even excitation compared to single excitation for infinite ground case.

## 3.4 Equivalent Circuit Modelling of Two Monopoles on a Small Platform    41

As in the case of a single monopole on a small platform, a series RLC resonator is used to model the monopole mode, while a parallel RLC resonator is used to model each of the chassis modes. The mutual coupling between the two monopoles is modeled as two cascaded T-section networks, while the coupling to the chassis modes is modeled as an L-section network. Fig.3.10 shows the equivalent circuit model of two monopoles on a 100 mm × 40 mm chassis. The resonant frequencies and the corresponding quality factors calculated in [11] are used as initial values for the resonator equivalent circuit. Using the ADS simulator, the susceptance or reactance slope parameters, the resonant frequencies and Q-factors of the monopole elements and chassis modes were optimized to achieve best agreement between EM simulation and circuit simulation. The equivalent circuit model in ADC of two monopoles at the middle of a 100 mm × 40 mm chassis short ends is shown in Appendix B. Good agreement between results for the monopole self and mutual impedances and reflection and transmission coefficients from the equivalent circuit model and EM simulation has been achieved as shown in Fig.3.11 for the antenna structure in Fig.3.7 with $d = 19.5$ mm. For this element position, only the 1st chassis mode will be excited because the monopoles are placed where the current distributions for the 2nd and 3rd chassis modes are at their maximum. Consequently, the equivalent circuit for this situation emphasizes the 1st chassis mode resonator (high susceptance slope) in order to correctly model the impedance variation, as seen in Fig.3.11. For other element positions, e.g., $d = 4$ mm as used in Fig.3.7 and Fig.3.9, both resonators may appear similar in admittance level.

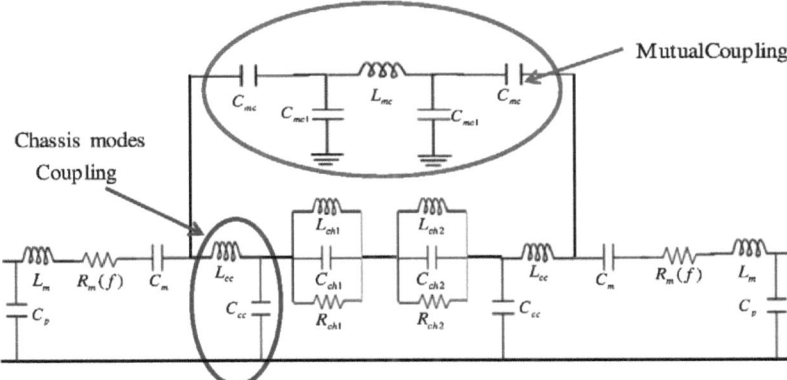

Figure 3.10: Equivalent circuit model of two monopoles placed at the middle of a 100 mm × 40 mm chassis short ends.

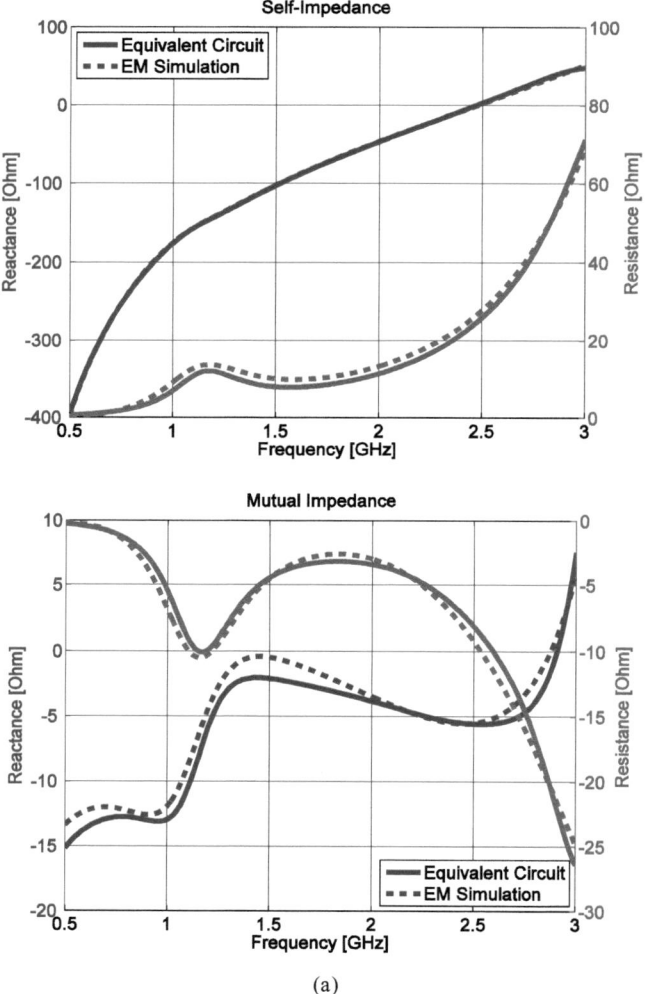

(a)

## 3.4 Equivalent Circuit Modelling of Two Monopoles on a Small Platform

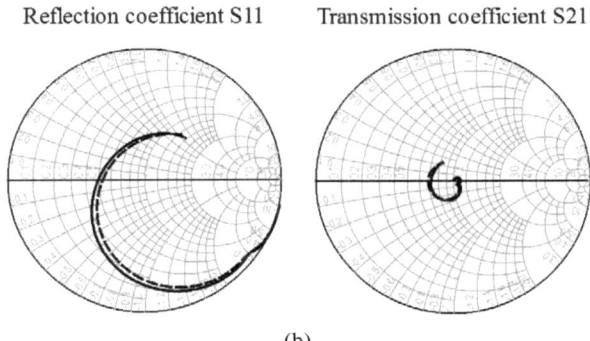

(b)

Figure 3.11: Equivalent circuit (solid) and EM simulation (dash). (a) Self- and mutual impedances and (b) reflection coefficient S11 and transmission coefficient S21 in Smith chart for two monopoles placed at $d = 19.5$ mm.

The resonator properties for the equivalent circuit model obtained from the optimization process are summarized in Tab. 3.2. As the electrical length of the chassis increases due to implementation of the monopole elements on the chassis, its resonant frequencies decrease compared to the unloaded case [44]. For $d = 19.5$ mm, the $2^{nd}$ chassis mode is nearly not excited because the monopoles are placed where the current distribution for the $2^{nd}$ mode is approximately at its maximum. This effect can be seen from the equivalent circuit as low resistance and high capacitance in the RLC circuit model for the $2^{nd}$ chassis mode, as indicated in Tab.3.2. This means that the corresponding mode is more probable to store energy rather than to radiate.

The equivalent circuit model can be used to separate the effect of chassis modes on the self- and mutual impedances and their contribution to the total radiation. Fig.3.12 shows the effect of the chassis modes on the self- and mutual impedances with calculations performed by the circuit model with either both chassis mode resonators effective or one of the chassis mode resonators short-circuited. Below the monopole quarter-wave resonance, the self-impedance for the case of both chassis mode resonators effective is nearly the same as with the $1^{st}$ chassis mode only (peak only seen at the chassis $1^{st}$ mode resonance). This verifies the excitation of the $1^{st}$ chassis mode, while the $2^{nd}$ chassis mode is approximately not excited. However, without the $1^{st}$ chassis mode resonator, the peak in the self-impedance real part cannot be modeled properly. On the other hand, the mutual impedance real part dips and the reactive part steeply slopes at the resonance of the 1$^{st}$ chassis mode. At frequencies above the quarter-wave resonance of the monopoles we observe that both resonators are required to model the impedance behavior correctly.

The percentage contribution of each chassis mode and monopole to the total radiation calculated by the equivalent circuit are shown in Fig.3.13. It can be seen that at the 1 GHz frequency range, 95% of the power is radiated by the chassis modes (mainly the 1$^{st}$ chassis mode), while at the 2.5 GHz frequency range, the 2$^{nd}$ chassis mode is less excited to contribute only about 10% of radiation power, same as the 1$^{st}$ mode. Thus, the contribution of the chassis modes to the total radiation doesn't exceed 20%, while the quarter-wavelength monopoles are the main radiators at the 2.5 GHz frequency range.

TABLE 3.2: RESONATOR PROPERTIES FOR LOADED AND UNLOADED CHASSIS.

| Chassis modes | Loaded chassis | | | | | Unloaded chassis | |
|---|---|---|---|---|---|---|---|
| | $C$ pF | $L$ nH | $R$ Ω | $f_r$ GHz | $Q$ | $f_r$ GHz | $Q$ |
| 1$^{st}$ chassis mode | 3.034 | 5.798 | 78.694 | 1.21 | 1.8 | 1.26 | 2.3 |
| 2$^{nd}$ chassis mode | 38 | 0.097 | 5.445 | 2.62 | 3.38 | 2.68 | 3 |

## 3.4 Equivalent Circuit Modelling of Two Monopoles on a Small Platform

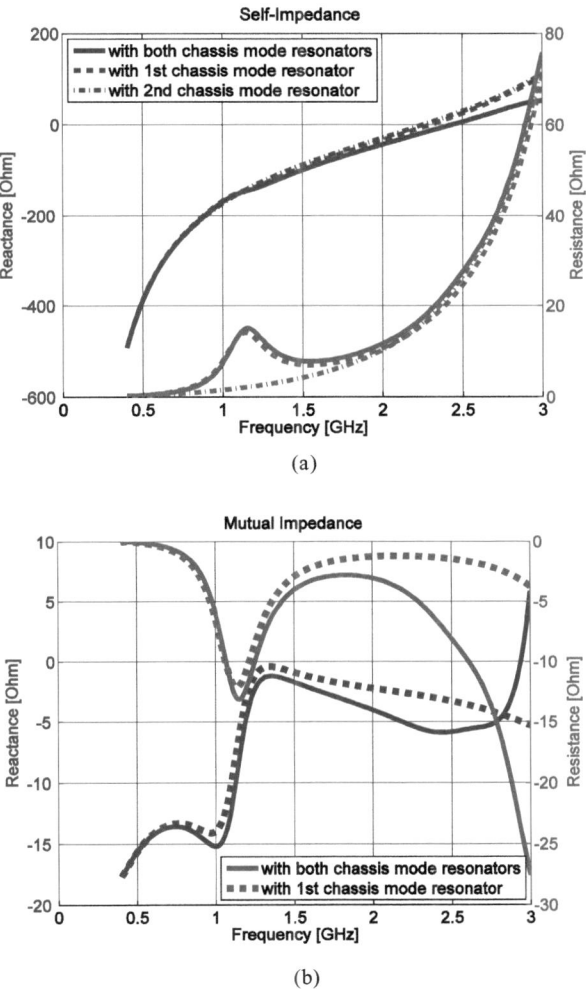

Figure 3.12: Effect of chassis modes on the (a) self-impedance and (b) mutual impedance of two monopoles placed at $d = 19.5$ mm.

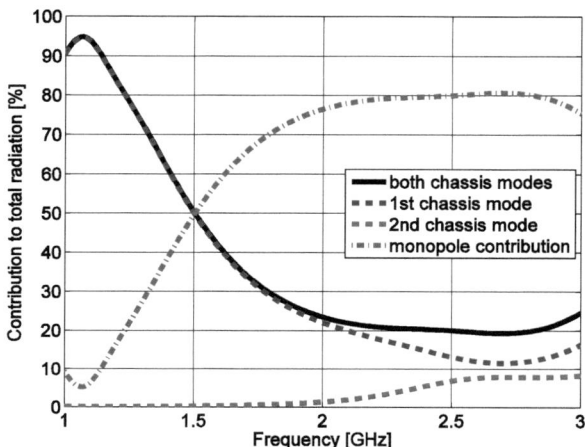

Figure 3.13: Contribution of chassis modes to the total radiated power for two monopoles placed at $d = 19.5$ mm of a 100 mm × 40 mm chassis.

For selective excitation of the $2^{nd}$ and $3^{rd}$ chassis modes, the two monopoles should be placed at the middle of the chassis long ends. The mutual coupling and chassis modes coupling in the equivalent circuit model of Fig.3.10 are modified to match for the new position of the monopoles. At this position, the $1^{st}$ chassis mode resonator will be short-circuited and a parallel RLC resonator should be added to model the $3^{rd}$ chassis mode. The mutual coupling and chassis modes coupling are modeled as T-section and L-section networks respectively. The parallel RLC resonators modeling the $2^{nd}$ and $3^{rd}$ chassis modes are rearranged such that the circuit properly works for sequential and all at once excitation settings of the ports.

For sequential excitation, one of the two ports is excited and the other is matched, while for all at once excitation, the two ports are excited simultaneously. The modified equivalent circuit model for two monopoles placed at the middle of a 100 mm × 40 mm chassis long ends is shown in Fig.3.14. The equivalent circuit model in ADC of two monopoles at the middle of a 100 mm × 40 mm chassis long ends is shown in Appendix C. Good agreement between the equivalent circuit model results and EM simulation can be seen in Fig.3.15. The resonator properties for the equivalent circuit model obtained from the optimization process are summarized in Tab. 3.3.

## 3.4 Equivalent Circuit Modelling of Two Monopoles on a Small Platform

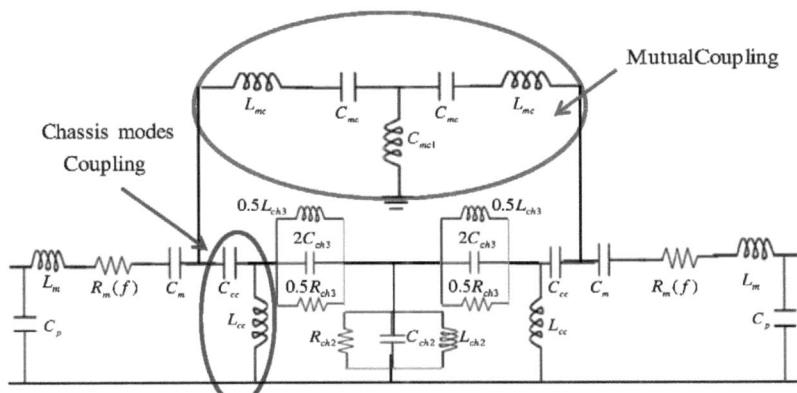

Figure 3.14: Equivalent circuit model of two monopoles placed at the middle of the chassis long ends.

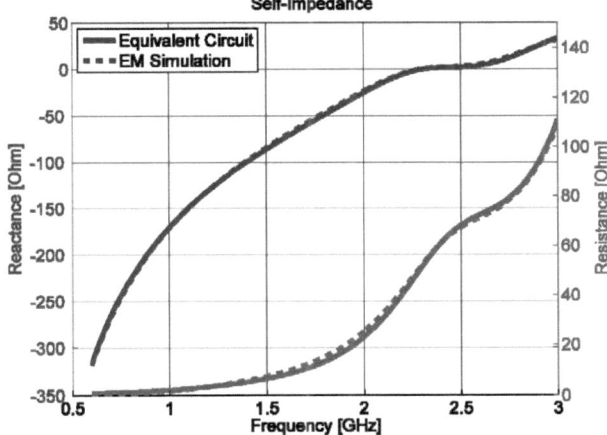

48   Ch3: Study of MC and Chassis Modes Coupling through the Equivalent Circuit Model of MPs on a Small Platform

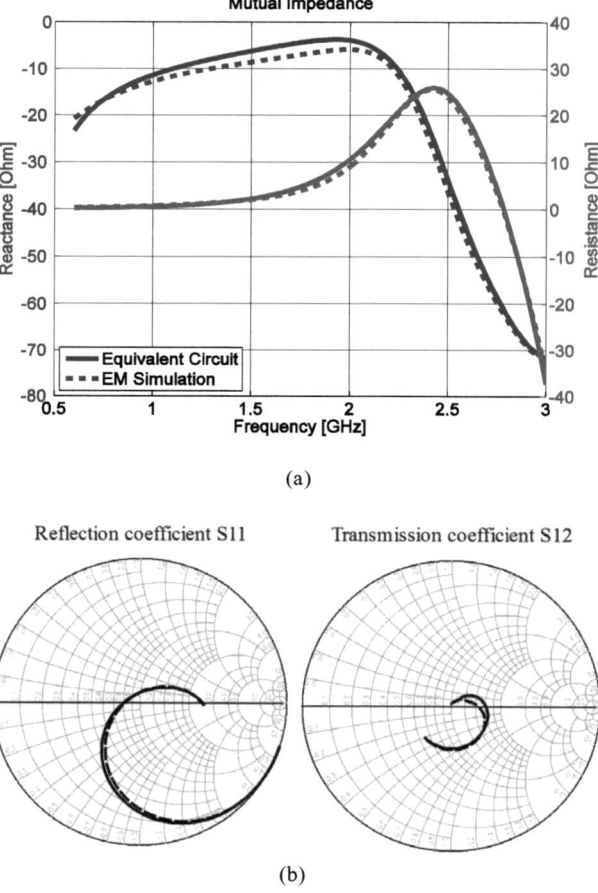

Figure 3.15: Equivalent circuit (solid) and EM simulation (dash). (a) Self- and mutual impedance and (b) reflection coefficient S11 and transmission coefficient S21 in Smith chart for two monopoles placed at the middle of a 100 mm × 40 mm chassis long ends.

## 3.4 Equivalent Circuit Modelling of Two Monopoles on a Small Platform

TABLE 3.3: RESONATOR PROPERTIES FOR LOADED AND UNLOADED CHASSIS.

| Chassis modes | Loaded chassis | | | | | Unloaded chassis | |
|---|---|---|---|---|---|---|---|
| | $C$ pF | $L$ nH | $R$ Ω | $f_r$ GHz | $Q$ | $f_r$ GHz | $Q$ |
| $2^{nd}$ chassis mode | 4.242 | 0.925 | 36.935 | 2.54 | 2.5 | 2.68 | 3 |
| $3^{rd}$ chassis mode | 4.516 | 0.763 | 26.004 | 2.71 | 2 | 2.74 | 2.5 |

The percentage contribution of each chassis modes and monopole to the total radiation calculated by the equivalent circuit are shown in Fig.3.16. It can be seen that at the 1 GHz frequency range, the monopole mode is the main radiator, while the contribution of chassis modes to the total radiation is approximately 25%. At the 2.5 GHz frequency range, the contribution of the $2^{nd}$ and $3^{rd}$ chassis modes reaches 50% of the total radiation.

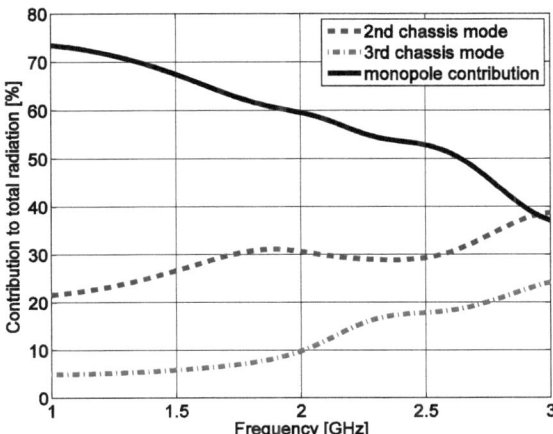

Figure 3.16: Contribution of chassis modes to the total radiated power for two monopoles placed at middle of a 100 mm × 40 mm chassis long ends.

# Chapter 4

# Analysis of Mutual Coupling and Chassis Modes Coupling in the Monopole Four Square Array Antenna (MFSAA)

## 4.1 Introduction

Wireless communication systems with omnidirectional antennas suffer from multipath and interference signals in indoor propagation environment. Multi-beam antennas and adaptive arrays are one of the main popular techniques used for interference suppression by enhancing the signal strength and forming nulls in the direction of interferences. In antenna arrays, mutual coupling plays a significant effect on the array performance particularly when a space less than half-wavelength is available between the antenna elements as in the case of mobile terminals. The problem of mutual coupling is critical in beamfoming antenna arrays. It affects significantly the depths and positions of the nulls towards the interferences. In [45-47], the mutual coupling effect on the performance of adaptive arrays and direction of arrival estimation has been addressed. In [48], different array analysis methods for pattern modelling including the effect of mutual coupling are presented and compared.

In the Monopole Four Square Array Antenna (MFSAA), four monopole antenna elements are placed on a square ground for beam scanning and sectorization of 360° azimuthal coverage. The beamforming function of the array is strongly affected by the mutual coupling between array elements and the chassis modes coupling between array elements and the ground. In this chapter, the MFSAA performance under the effect of mutual coupling and chassis modes coupling is presented and discussed.

## 4.2 Monopole Four Square Array Antenna (MFSAA)

In the MFSAA, four monopole antennas are placed in a square arrangement on a 100 mm × 100 mm chassis and fed in order to generate four overlapping beams in azimuth. Each of the monopole elements has a quarter-wavelength length $L$ at a resonant frequency of 2.45 GHz and a radius $r$. The distance between the adjacent array elements is $d$. The geometry of the MFSAA is shown in Fig.4.1.

### 4.2.1 Radiation Characteristics of the MFSAA

For the MFSAA in Fig.4.1, the total radiation intensity under the assumption that the array elements are isolated (the mutual coupling effect is not considered) can be expressed as [49]

$$F(\theta,\phi) = F_{Mp}(\theta)|AF(\theta,\phi)|^2, \qquad (4.1)$$

where $F_{Mp}(\theta)$ is the radiation intensity of a single monopole antenna, and $AF(\theta,\phi)$ is the array factor. The radiation pattern for a single monopole is shown in Fig.4.2. The array factor $AF(\theta,\phi)$ is a function of the distance $d$ between the array elements and the excitation vector for the array. For a $M \times N$ planar array with a uniform amplitude excitation, the normalized array factor can be written as [49]

$$AF(\theta,\phi) = \frac{1}{M}\frac{\sin\left(\frac{M}{2}\psi_x\right)}{\sin\left(\frac{\psi_x}{2}\right)}\frac{1}{N}\frac{\sin\left(\frac{N}{2}\psi_y\right)}{\sin\left(\frac{\psi_y}{2}\right)}, \qquad (4.2)$$

where

$$\begin{aligned}\psi_x &= kd_x\sin\theta\cos\phi + \beta_x \\ \psi_y &= kd_y\sin\theta\sin\phi + \beta_y,\end{aligned} \qquad (4.3)$$

where $k$ is the propagation constant, $d_x$ and $\beta_x$ are the spacing and phase shift of excitation in x-direction respectively, and the spacing and phase shift of excitation in y-direction are presented respectively by $d_y$ and $\beta_y$.

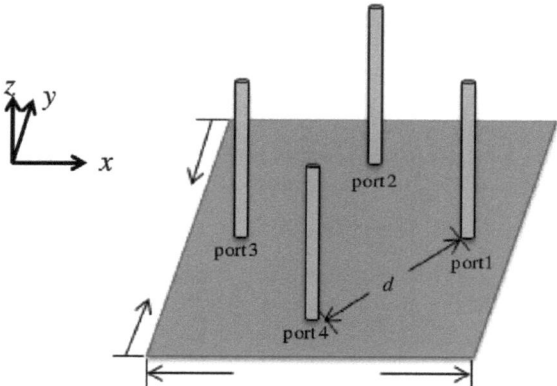

Figure 4.1: Monopole four square array antenna (MFSAA) with monopole elements (length $L = 0.25\lambda$, radius $r = 0.01\lambda$) and element spacing $d = 0.35\lambda$.

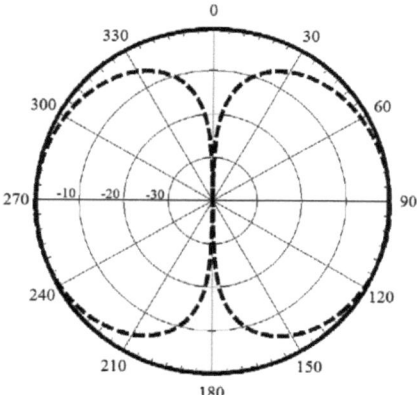

Figure 4.2: Azimuth (solid) and elevation (dash) radiation pattern of a quarter-wavelength monopole at 2.45 GHz.

## 4.2 Monopole Four Square Array Antenna

For the MFSAA in Fig.4.1, $M = N = 2$, $\beta_x = \beta_y$, and $d_x = d_y = d$. The excitation vector of the MFSAA for a beamforming in $\phi = 45°$ direction and nulls in $\phi = 135°, 225°,$ and $315°$ directions is chosen to be uniform in amplitude and with $\beta_x = \beta_y = 90°$ for $d = 0.35\lambda$ as given below for the antenna elements 1, 2, 3, and 4 in Fig.4.1 respectively.

$$\begin{bmatrix} V_{o1} \\ V_{o2} \\ V_{o3} \\ V_{o4} \end{bmatrix} = \begin{bmatrix} 1 \\ e^{j\pi/2} \\ -1 \\ e^{j\pi/2} \end{bmatrix} \quad (4.4)$$

The radiation pattern of the MFSAA azimuth beam with a monopole length $L = 0.25\lambda$ at 2.45 GHz, radius $r = 0.01\lambda$, and varied distance $d$ between adjacent array elements is shown in Tab.4.1. For $d = 0.35\lambda$, an azimuth beam in $\phi = 45°$ direction and nulls in $\phi = 135°, 225°,$ and $315°$ directions is obtained. For the MFSAA in Fig.4.1, where the monopole elements are placed at the middle of the chassis ends, an azimuth beam in $\phi = 0°$ direction is obtained. For the monopole elements placed at the corners of the chassis (rotating by $45°$), the azimuth beam becomes in $\phi = 45°$ direction. For both configurations the three other beams are generated by cyclic rotation of the port excitation.

### 4.2.2 Directivity of the MFSAA

The directivity of an array antenna defined as the ratio of the radiation intensity in a given direction from the array to the radiation intensity averaged over all directions [49], it can be written as

$$D(\theta, \phi) = 4\pi \frac{F(\theta, \phi)}{\int_0^{2\pi} \int_0^{\pi} F(\theta, \phi) \sin\theta \, d\theta \, d\phi}, \quad (4.5)$$

where $F(\theta, \phi)$ is the total radiation intensity defined by (4.1). If the direction is not specified, the direction of maximum $F(\theta, \phi)$ (maximum directivity $D_0$) is implied and given by

$$D_0 = 4\pi \frac{F(\theta,\phi)|_{max}}{\int_0^{2\pi}\int_0^{\pi} F(\theta,\phi)\sin\theta \, d\theta \, d\phi} \qquad (4.6)$$

### 4.2.3 Front-to-Back Ratio (F/B) of the MFSAA

The front-to-back ratio is the difference between the maximum directivity in the forward direction and the backward direction under 180°. The directivity and (F/B) ratio of the MFSAA azimuth beam in xy-plane with a monopole length $L = 0.25\lambda$ at 2.45 GHz, radius $r = 0.01\lambda$, and varied distance $d$ between adjacent array elements is shown in Tab. 4.1. For $d = 0.35\lambda$, the desired azimuth beam in $\phi = 45°$ direction and nulls in $\phi = 135°, 225°,$ and $315°$ directions is obtained with a high front-to-back ratio (72.21 dB) and a directivity of 8.94 dB. Side lobes level (SL) of less than -10 dB is obtained.

### 4.2.4 Maximum Absolute Gain of the MFSAA

The antenna absolute gain is related to the directivity, it takes into account the losses at the input terminals (mismatch losses) in addition to the conduction and dielectric losses of the structure (internal losses). Thus the maximum absolute gain can be defined as [49]

$$G_{0abs} = \eta_r \eta_m D_0, \qquad (4.7)$$

where $\eta_r$ and $\eta_m$ are the radiation efficiency and mismatch efficiency respectively of the MFSAA with length $L$ and radius $r$. $D_0$ is the maximum directivity defined in (4.6). The radiation efficiency $\eta_r$ is given by

$$\eta_r = \frac{R_{rad}}{R_{rad} + R_{dis}}, \qquad (4.8)$$

$R_{rad}$ and $R_{dis}$ are the monopole radiation and dissipation resistances respectively. The radiation resistance can be approximated for a thin monopole as [50]

$$R_{rad} = \frac{30}{\sin^2(kL)} \int_0^{\pi} \frac{[\cos(kL\cos\theta) - \cos(kL)]^2}{\sin\theta} d\theta, \qquad (4.9)$$

while the dissipation resistance is approximated by [51]

## 4.2 Monopole Four Square Array Antenna

TABLE 4.1: AZIMUTH BEAM PATTERN, DIRECTIVITY, AND (F/B) OF THE MFSAA IN FIG.4.1 FOR EXCITATION VECTOR IN (4.4) AND VARIED SEPARATION BETWEEN ADJACENT ANTENNA ELEMENTS $d$ .(MUTUAL COUPLING AND CHASSIS MODES COUPLING ARE NOT CONSIDERED)

| $d/\lambda$ | Directivity in xy-plane dB | (F/B) Ratio dB | Azimuth Beam Pattern |
|---|---|---|---|
| 0.30 | 8.03 | 25.11 | |
| 0.35 | 8.94 | 72.21 | |
| 0.40 | 9.45 | 27.54 | |

$$R_{dis} = \frac{1}{2rk\sin^2(kL)}\left(\frac{kL}{2} - \frac{\sin(2kL)}{4}\right). \tag{4.10}$$

The mismatch efficiency $\eta_m$ is given by

$$\eta_m = 1 - |\Gamma|^2, \tag{4.11}$$

with $\Gamma$ the voltage reflection coefficient at the input terminals of the antenna defined as $(Z_{in} - Z_o)/(Z_{in} + Z_o)$, where $Z_{in}$ is the antenna input impedance and $Z_o$ is the source impedance (50 $\Omega$ impedance). If the antenna is matched ($|\Gamma| = 0$), then the antenna absolute gain $G_{0abs}$ will be equal to the antenna gain $G = \eta_r D_0$.

## 4.3 Mutual Coupling in the MFSAA

For simplicity, the mutual coupling between a two-element array will be considered for both transmit and receive modes. The same concept can be applied for the MFSAA. In a transmit mode, the two array elements in Fig.4.3 are driven simultaneously. The excited current in the 1$^{st}$ element radiates energy in free space. Part of the incident energy is coupled to the 2$^{nd}$ element and causes an induced current in that element. The induced current in the 2$^{nd}$ element radiates energy in free space. Part of the energy is coupled back to the 1$^{st}$ element causes an induced current in that element. The induced current caused by the 2$^{nd}$ element modifies the current in the 1$^{st}$ element which in turns modifies the input impedance of the 1$^{st}$ element. As a result, the radiation of the 1$^{st}$ element will be a combination of both antenna elements 1 and 2. In Fig.4.3, the plot shows that antenna1 is driven, same plot is applied for antenna2 of a two-element array in a transmit mode.

By equations, the effect of the mutual coupling can be also indicated. In Fig.4.3, the port terminal voltages $[V]$ relate to the port terminal currents $[I]$ according to

$$\begin{bmatrix} V_1 \\ V_2 \end{bmatrix} = \begin{bmatrix} Z_{11} & Z_{12} \\ Z_{21} & Z_{22} \end{bmatrix} \begin{bmatrix} I_1 \\ I_2 \end{bmatrix}. \tag{4.12}$$

From (4.12), the port terminal voltages can be expressed as

$$V_1 = Z_{11}I_1 + Z_{12}I_2 \tag{4.13}$$

$$V_2 = Z_{21}I_1 + Z_{22}I_2. \tag{4.14}$$

## 4.3 Mutual Coupling in the MFSAA

From Fig.4.3, the port terminal voltages relate to the port open source voltages by

$$V_1 = V_{o1} - Z_g I_1 \qquad (4.15)$$

$$V_2 = V_{o2} - Z_g I_2 \qquad (4.16)$$

where $Z_g$ is the source impedance and $V_{o1}$ and $V_{o2}$ are the open-circuit source voltages for antenna1 and antenna2, respectively. By substituting (4.16) in (4.14) and solve for $I_2$

$$I_2 = \frac{V_{o2} - Z_{21} I_1}{Z_{22} + Z_g}. \qquad (4.17)$$

Substituting (4.17) in (4.13) and rearranging (4.13) to get

$$V_1 = \left( Z_{11} - \frac{Z_{21} Z_{12}}{Z_{22} + Z_g} \right) I_1 + Z_{12} \frac{V_{o2}}{Z_{22} + Z_g}. \qquad (4.18)$$

The mutual coupling modifies the input impedance $Z_{11}$ at port1 by the factor $Z_{21} Z_{12}/(Z_{22} + Z_g)$, and accordingly the terminal voltage $V_1$ is modified resulting in a modified array pattern.

For a two-element array in receive mode as Fig.4.4 shows, the wave received by the 1st element causes current to flow in that element which causes energy radiation. Part of the energy is coupled to the 2nd element causing an induced current in that element, then energy radiation from the 2nd element and part of the energy is coupled back to the 1st element. The current in the 1st element is modified and accordingly the input impedance is also modified resulting in a modified radiation characteristics.

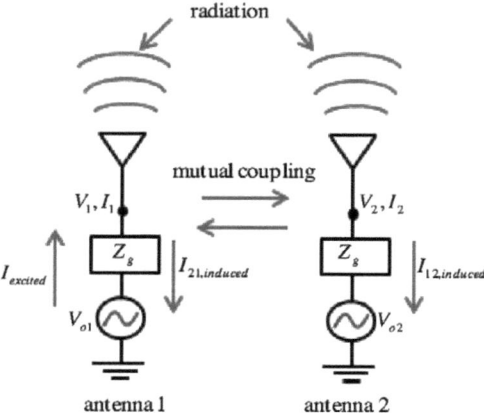

Figure 4.3: Model of a two-element array in a transmit mode.

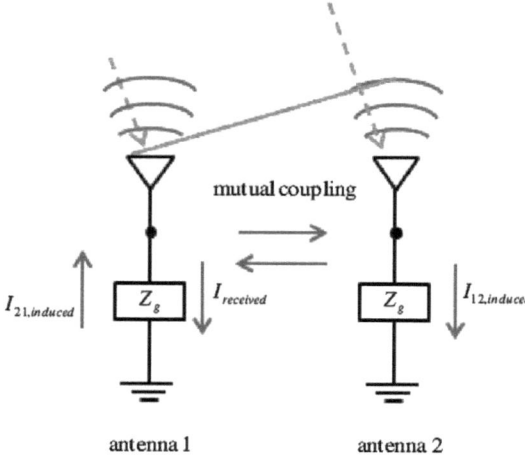

Figure 4.4: Model of a two-element array in a receive mode.

### 4.3.1 Mutual Coupling Effect on the MFSAA Radiation Pattern

The effect of mutual coupling and chassis modes coupling on the azimuth patterns of the MFSAA in $\phi = 45°$ direction shown in Tab.4.1 has been neglected. To see the effect of mutual coupling, the MFSAA on an infinite ground plane with $d = 0.35\lambda$ is considered. From EM simulation using Empire Xccel, as indicated in Fig.4.5, the mutual coupling between array elements fills the nulls at $\phi = 135°, 225°,$ and $315°$, and shifts the nulls at $\phi = 135°,$ and $315°$ by 15 degrees compared to the isolated elements pattern. A comparison between the pattern of the isolated and coupled elements is summarized in Tab.4.2 with respect to the beamwidth (BW), side lobes level (SL), beam crossover level (BC), and front-to-back ratio (F/B) for the MFSAA with $d = 0.35\lambda$. Due to coupling, the beamwidth becomes narrower, the side lobes level becomes higher, and the beam crossover level and the front-to-back ratio become lower compared to the isolated elements pattern. In the isolated elements pattern the effect of mutual coupling is neglected, while in the coupled elements pattern the effect of mutual coupling is considered.

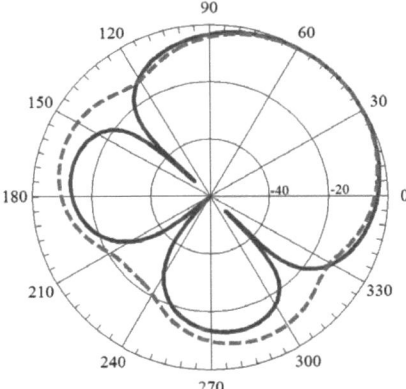

Figure 4.5: Isolated elements azimuth pattern (solid) and coupled elements azimuth pattern (dash) of the MFSAA.

TABLE 4.2: COMPARISON BETWEEN PATTERNS OF ISOLATED AND COUPLED ELEMENTS OF THE MFSAA.

|  | Isolated Pattern | Coupled Pattern |
|---|---|---|
| **Beamwidth** | 85° | 76° |
| **Side lobes level** | -12.47 dB | -7.87 dB |
| **Front-to-back ratio** | 72.21 dB | 22.71 dB |
| **Beam crossover level** | -3.42 dB | -4.3 dB |

## 4.4 Chassis Modes Coupling in the MFSAA

In addition to the effect of mutual coupling between array elements, the chassis modes coupling, as discussed in the second and third chapters, will affect the MFSAA input impedance and radiation pattern. For optimal coupling to the chassis modes, the exciter location has to be properly chosen. In [16, 17], the maximum bandwidth potential was obtained when the coupling element is placed at the corner of the chassis.

A quarter-wavelength monopole is placed at the corner of a 100 mm × 100 mm chassis for optimal excitation to the chassis dominant wavemodes within the frequency range from 0.5 GHz to 3 GHz. In [40, 42], the equivalent circuit model of one and two monopoles on a 100 mm × 40 mm chassis shows that the behavior of the monopole input resistance is strongly influenced by the chassis, while the monopole input reactance is dominated by the exciter which in this case is a monopole element.

The input resistance of a quarter-wavelength monopole placed at the corner of a 100 mm × 100 mm chassis is simulated using Empire Xccel and compared to the case of infinite ground plane as can be seen in Fig.4.6. The $1^{st}$, $2^{nd}$, and $3^{rd}$ chassis modes are seen as a peak in the monopole input resistance compared to the case of infinite ground. The resonant frequencies of the $1^{st}$, $2^{nd}$ and $3^{rd}$ chassis modes occur at 1.0, 1.64, and 2.85

## 4.4 Chassis Modes Coupling in the MFSAA

GHz, respectively. Due to loading of the chassis by a quarter-wavelength monopole, the resonant frequencies of the chassis modes are slightly tuned downward [44].

The 1$^{st}$ chassis mode is the half-wavelength mode of the chassis end, it is strongly coupled when the monopole is placed at the middle of the chassis end perpendicular to that end. In case of a single monopole placed at the corner of the chassis, the 1$^{st}$ chassis mode becomes the half-wavelength mode of the chassis diagonal. The current distribution from EM simulation using Empire Xccel for the 1$^{st}$ chassis mode at 1 GHz is shown in Fig.4.7 for different positions of a single monopole on a 100 mm × 100 mm chassis.

A quarter-wavelength monopole is placed at the corner of the chassis at 1.64 GHz to excite the 2$^{nd}$ chassis mode, the full-wavelength mode of the chassis diagonal. While the 3$^{rd}$ chassis mode (the full-wavelength mode of the chassis end) is excited at 2.85 GHz by two in-phase quarter-wavelength monopoles placed at the middle of the opposite chassis ends. The current distributions using Empire Xccel for the 2$^{nd}$ and 3$^{rd}$ chassis modes are shown in Fig.4.8a and Fig.4.8b respectively.

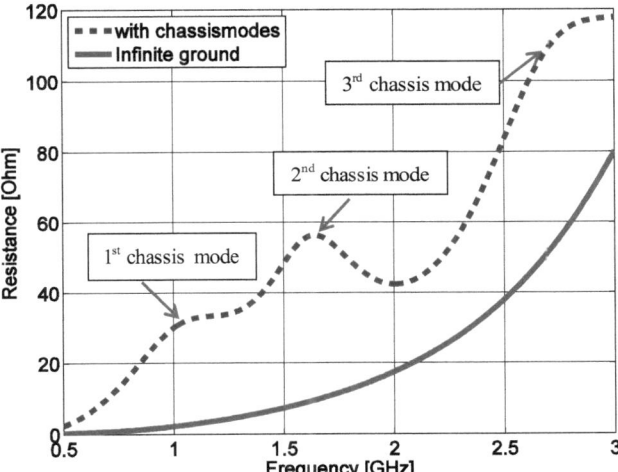

Figure 4.6: Input resistance of a single monopole placed at the corner of a 100 mm × 100 mm chassis.

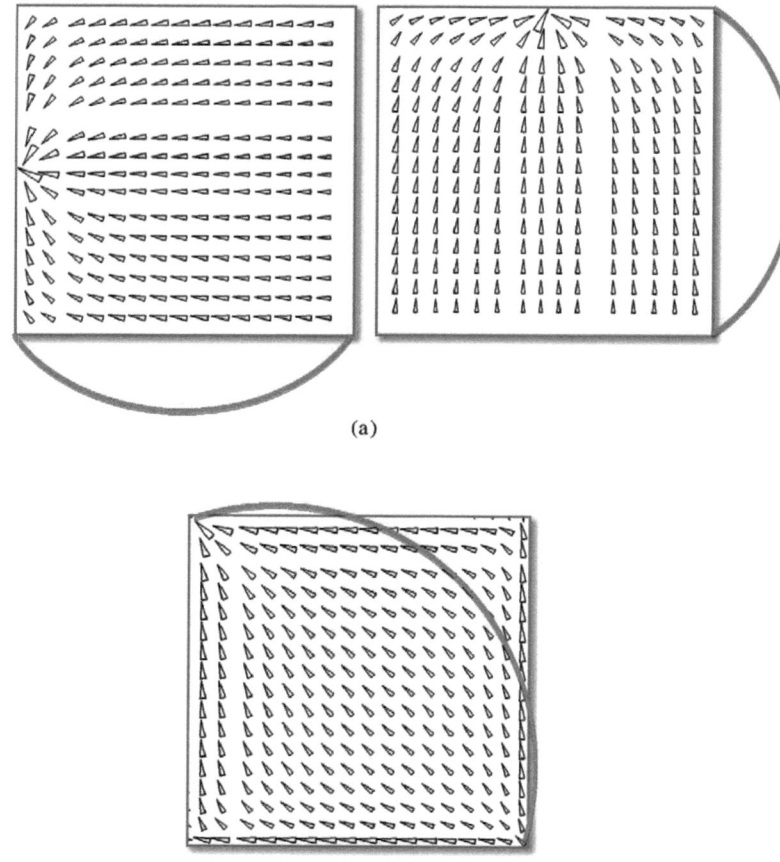

Figure 4.7: Current distribution of the half-wavelength mode at 1 GHz for a single monopole placed (a) at the middle of the chassis end, and (b) at the corner of the chassis. (Arrows indicate the current direction)

## 4.4 Chassis Modes Coupling in the MFSAA

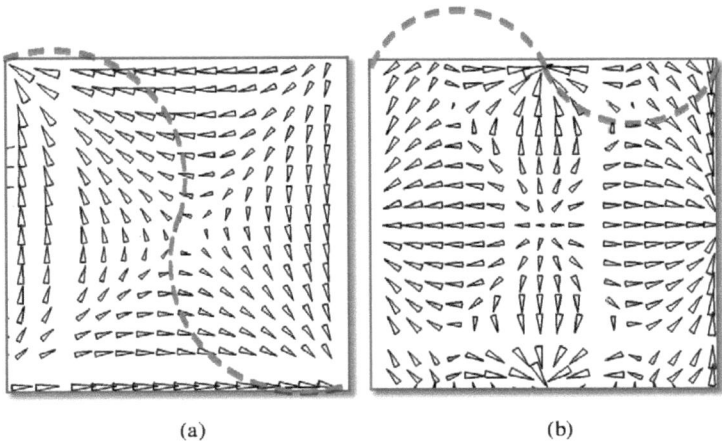

(a)  (b)

Figure 4.8: Current distribution of (a) the $2^{nd}$ chassis mode at 1.64 GHz, and (b) the $3^{rd}$ chassis mode at 2.85 GHz. (Arrows indicate the current direction)

### 4.4.1 Chassis Modes Coupling Effect on the MFSAA Radiation Pattern

To study the effect of chassis modes coupling, two configurations of the MFSAA on a 100 mm × 100 mm chassis are presented. The first one is the MFSAA placed at the corners of the chassis, while the second one is the MFSAA placed at the middle of the chassis ends. For the MFSAA placed at the chassis corners with the excitation vector of (4.4), from EM simulation using Empire Xccel, the azimuth pattern is similar to that for the infinite ground plane with slightly higher side lobes level as shown in Fig.4.9a, while in elevation pattern, the nulls are filled and the pattern is distorted compared to the infinite ground as Fig.4.9b shows.

The arrangement of the MFSAA at the corners of the chassis with $d = 0.35\lambda$ enables the excitation of the $1^{st}$ and $2^{nd}$ chassis modes. The resonant frequencies of those two modes occur at 1 and 1.64 GHz respectively. The radiation patterns for those two modes are plotted using Empire Xccel in Fig.4.10. Two short probes are placed at the most far corners of a 100 mm × 100 mm chassis with 180° out of phase at 1 GHz to excite the $1^{st}$ chassis mode (half-wavelength mode of chassis diagonal) and in-phase at 1.64 GHz to excite the $2^{nd}$ chassis mode (full-wavelength mode of chassis diagonal). The E-plane patterns of the $1^{st}$ and $2^{nd}$ chassis modes presented in Fig4.10 explain the less effect of chassis modes on the azimuth pattern of the MFSAA placed at the corners of a 100 mm × 100 mm chassis. While the H-plane pattern of the $1^{st}$ and $2^{nd}$ chassis modes presented in Fig4.10 is responsible for the distortion in the elevation pattern of the MFSAA placed at the corners of a 100 mm × 100 mm chassis.

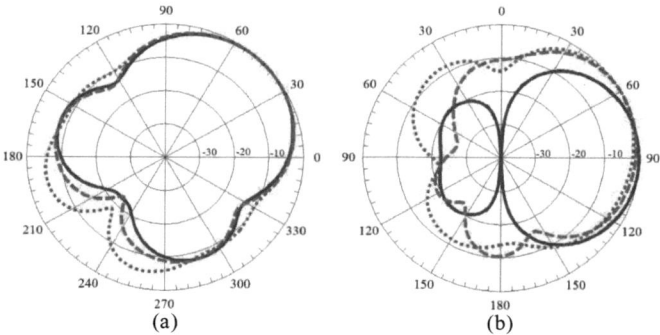

Figure 4.9: (a) Azimuth and (b) elevation patterns at phi = 0° with mutual coupling (solid), mutual and chassis modes coupling of the MFSAA at the chassis corners (dash), and mutual and chassis modes coupling of the MFSAA at the middle of the chassis ends (dot).

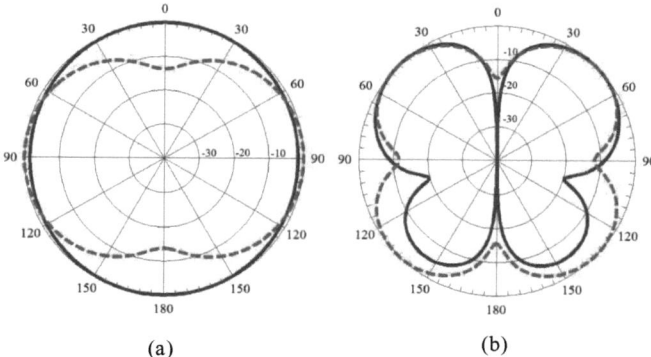

Figure 4.10: H-plane (solid) and E-plane (dash) patterns of (a) the $1^{st}$ chassis mode at 1GHz and (b) the $2^{nd}$ chassis mode at 1.64 GHz.

## 4.4 Chassis Modes Coupling in the MFSAA

For the MFSAA placed at the middle of the chassis ends, the nulls in azimuth pattern are more filled and the side lobes level is higher due to strong excitation to the 3$^{rd}$ chassis mode as Fig.4.9a shows. The radiation pattern of the 3$^{rd}$ chassis mode is plotted in Fig.4.11. Two short probes are placed at the middle of the opposite ends of a 100 mm × 100 mm chassis and excited in-phase at 2.85 GHz to excite the 3$^{rd}$ chassis mode (the full-wavelength mode of the chassis end).

The H-plane pattern of the 3$^{rd}$ chassis mode shown in Fig.4.11 distorts the elevation pattern of the MFSAA placed at the middle of a 100 mm × 100 mm chassis ends. While the E-plane pattern of the 3$^{rd}$ chassis mode is responsible for more filling of nulls and higher side lobes of the MFSAA azimuth pattern compared to the case of MFSAA placed at the chassis corners.

For comparison, the current distribution at 2.45 GHz for both configurations of the MFSAA placed at a 100 mm × 100 mm chassis is shown in Fig.4.12. As noticed in Fig.4.12a, the 1$^{st}$ chassis mode is excited by the two out-of-phase array elements, while the 2$^{nd}$ chassis mode is excited by the other two in-phase array elements. The two in-phase array elements in Fig.4.12b excite the 3$^{rd}$ chassis mode.

In above discussion and later on, E-plane and H-plane are used for the chassis modes patterns, while azimuth and elevation patterns are used for the MFSAA patterns. E-plane for the chassis modes is the xy-plane in Fig.4.1, and yz-plane is the H-plane.

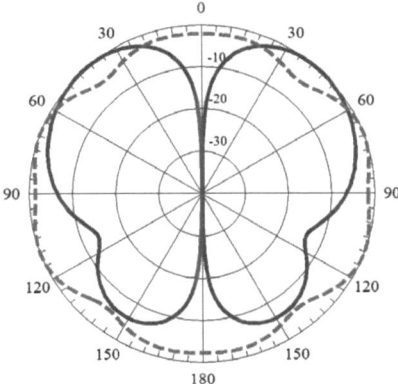

Figure 4.11: H-plane (solid) and E-plane (dash) patterns of the 3$^{rd}$ chassis mode at 2.85GHz.

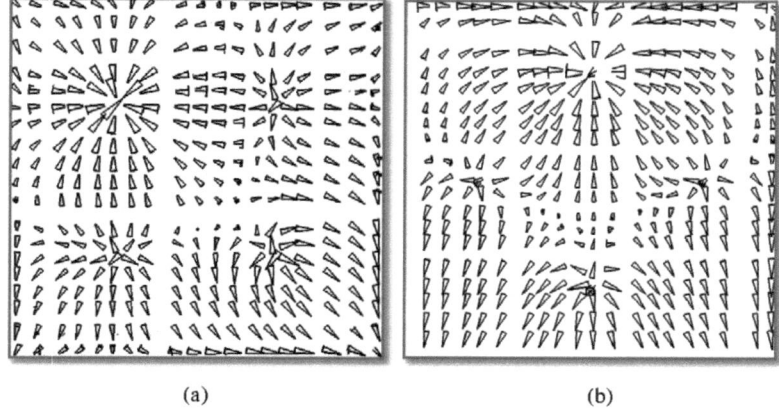

(a)                                (b)

Figure 4.12: Current distribution at 2.45 GHz of the MFSAA placed (a) at the chassis corners and (b) at the middle of the chassis ends. (Arrows indicate the current direction).

# Chapter 5

## Compensating for Mutual Coupling and Chassis Modes Coupling in the MFSAA

### 5.1 Introduction

The radiation pattern of the MFSAA is strongly affected by the mutual coupling and chassis modes coupling as discussed in the previous chapter. In particular, the depths and positions of the nulls are affected by both mutual coupling and chassis modes coupling. Many attempts have been made to compensate for the effect of mutual coupling in receiving antenna arrays for direction finding and interference suppression [52-54] while some authors try to compensate by decoupling the input ports of the feeding network [55-59]. In [60], the effect of ground parameters on array mutual coupling for direction of arrival estimation is investigated, when the skin depth of the ground plane increases or the loss tangent decreases, the mutual impedance decreases. For the MFSAA, to compensate for the effects of mutual coupling, the case of an infinite ground plane will be considered first, and later the effect of chassis modes will be taken into account. Different decoupling networks are proposed and compared for both port- and pattern-isolation of the MFSAA at 2.45 GHz. Coupled patterns will be used to point to coupled radiators. To compensate for coupled patterns in multi-beam antenna for the purpose of forming nulls in the direction of interferences, the feed network has to be optimized. Modifying the feed network keeps the ports and then the patterns correlated. Design of decoupling networks becomes of essential to get decoupled ports and then decoupled patterns.

## 5.2   Compensating for Coupled Radiation Patterns of the MFSAA

As discussed in ch.4, the mutual coupling in the MFSAA modifies the monopole input impedances, and accordingly the terminal voltages are modified resulting in a modified array pattern. The distortion of the radiation pattern of the MFSAA for the purpose of beamforming can be compensated by modifying the feeding network of the MFSAA. In [61], an optimization of the MFSAA using the method of genetic algorithm is performed including modification of the array geometry and the feeding network for the purpose of beamforming. Optimization with genetic algorithm requires long simulation time. In addition to guarantee convergence, the initial values of the parameters to be optimized have to be properly chosen. In the following, a more straight forward method will be presented: The mutual coupling effect in the MFSAA radiation pattern is compensated based on the active input impedance of the array elements. For simplicity, let's consider the case of a two-element array. The port terminal voltages $[V]$ relates to the port terminal currents $[I]$ according to

$$\begin{bmatrix} V_1 \\ V_2 \end{bmatrix} = \begin{bmatrix} Z_{11} & Z_{12} \\ Z_{21} & Z_{22} \end{bmatrix} \begin{bmatrix} I_1 \\ I_2 \end{bmatrix}. \tag{5.1}$$

From (5.1), the port terminal voltages can be expressed as

$$V_1 = Z_{11}I_1 + Z_{12}I_2 \tag{5.2}$$

$$V_2 = Z_{21}I_1 + Z_{22}I_2. \tag{5.3}$$

In Fig.4.3, the port open source voltages relate to the port terminal voltages by

$$V_{o1} = V_1 + Z_g I_1 \tag{5.4}$$

$$V_{o2} = V_2 + Z_g I_2, \tag{5.5}$$

where $Z_g$ is the source impedance (50 $\Omega$ impedance). $V_{o1}$ and $V_{o2}$ are the open source voltages for antenna1 and antenna2, respectively. Due to mutual coupling, the port terminal voltages $V_1$ and $V_2$ are modified by the terms $Z_{12}I_2$ and $Z_{21}I_1$, respectively,

## 5.2 Compensating for Coupled Radiation Patterns of the MFSAA

which means that to compensate for coupling, the port open source voltages $V_{o1}$ and $V_{o2}$ have to be modified by the same terms. The modified open source voltages are

$$V'_{o1} = V_{o1} + Z_{12}I_2 \qquad (5.6)$$

$$V'_{o2} = V_{o2} + Z_{21}I_1. \qquad (5.7)$$

The port terminal currents $I_1$ and $I_2$ in (5.6) and (5.7) are given by

$$I_1 = \frac{V_{o1}}{Z_g + Z_{in1}} \qquad (5.8)$$

$$I_2 = \frac{V_{o2}}{Z_g + Z_{in2}}, \qquad (5.9)$$

where, $Z_{in1}$ and $Z_{in2}$ are the input impedances seen looking into the antenna terminals when one element is excited and the other is matched to the 50 $\Omega$. Due to mutual coupling, the modified feeding network for beamforming depends now on the impedance matrix $[Z]$. The conventional definition of the impedance matrix elements, however, is not satisfactory for our present application since the open-circuit assumption involved in the calculation of the impedance elements does not take into account the scattering effect from antenna elements in the array when they are open-circuited. This effect is modelled as a parasitic capacitor $C_p$ in the equivalent circuit model, Fig.3.2. This shortcoming can be improved by analyzing a transmit array, where all array elements are active. The input impedances will be denoted as active input impedances to point to antenna array in a transmit mode. From (5.2) and (5.3), the active input impedances are given by

$$Z_{active1} = Z_{11} + C_{12}\frac{I_2}{I_1} \qquad (5.10)$$

$$Z_{active2} = Z_{22} + C_{21}\frac{I_1}{I_2}. \qquad (5.11)$$

Unlike in the conventional definition, the magnitudes and phases of both antenna array terminal currents will be considered in the calculation of mutual coupling. Thus, the mutual impedances $Z_{12}$ and $Z_{21}$ from the conventional definition will be replaced by the coupling coefficients $C_{12}$ and $C_{21}$, respectively. $Z_{11}$ and $Z_{22}$ are the antenna element self-impedances. Active input impedances in (5.10) and (5.11) can be calculated from EM simulation using Empire Xccel. By substituting (5.8) and (5.9) in (5.10) and (5.11) and solve for $C_{12}$ and $C_{21}$, different values for coupling coefficients $C_{12}$ and $C_{21}$ are obtained as compared to the conventional method. Finally, the modification terms in (5.6) and (5.7) can be calculated now to derive the new feed network to compensate for mutual coupling effect. The same work is extended for the MFSAA in Fig.4.1. The monopole elements are a quarter-wavelength monopoles at 2.45 GHz, the monopole radius $r$ is $0.01\lambda$, and the separation distance $d$ between adjacent array elements is $0.36\lambda$ while the separation distance between nonadjacent array elements is $0.51\lambda$. The active input impedance of the MFSAA ports is given as

$$Z_{activej} = Z_{jj} + \sum_{i=1}^{4} C_{ji} \frac{I_i}{I_j}, \qquad (5.12)$$

where $(j = 1,2,...,4)$, $i = (1,2,...,4)$, $j \neq i$. The excitation of the MFSAA without coupling for a beamforming in $\phi = 45°$ direction and nulls in $\phi = 135°, 225°,$ and $315°$ directions is given by

$$\begin{bmatrix} V_{o1} \\ V_{o2} \\ V_{o3} \\ V_{o4} \end{bmatrix} = \begin{bmatrix} 1 \\ e^{j\pi/2} \\ -1 \\ e^{j\pi/2} \end{bmatrix} \qquad (5.13)$$

The modified open source voltages $V'_{oj}$ for $(j=1,2,...,4)$ of the MFSAA with coupled elements are equal to

$$V'_{oj} = V_{oj} + \sum_{i=1}^{4} C_{ji} I_i, \; i \neq j, \qquad (5.14)$$

where the port terminal currents $I_i$ for $(i = 1,2,...,4)$ are equal to

$$I_i = \frac{V_{oi}}{Z_g + Z_{activei}}. \qquad (5.15)$$

The active input impedances $Z_{activej}$ for $(j=1,2,...,4)$ in (5.12) are calculated from EM simulation by the help of Empire Xccel. The mutual coupling coefficients are calculated

## 5.2 Compensating for Coupled Radiation Patterns of the MFSAA 71

from the input active impedances for the excitation in (5.13) and compared with those calculated from the conventional method. The mutual impedances from both methods are plotted in Fig.5.1. The resistance behavior of $Z_{12}, C_{12}$ and $Z_{13}, C_{13}$ from both methods is nearly the same up to the resonant frequency of the monopoles (2.45 GHz). With increasing the frequency (above 2.5 GHz), the difference between mutual resistances from the two methods increases as the array elements electrical size becomes comparable to the wavelength where the self-impedance steeply increases (small terminal current) and the scattering effect rises due to increasing current at the monopole conductor center. The reactance behavior of $Z_{12}, C_{12}$ and $Z_{13}, C_{13}$ is close for nonadjacent array elements from both methods, while for adjacent array elements, the difference between mutual reactances increases with the frequency increase. The monopole elements in the MFSAA are identical, this implies that $Z_{ii} = Z_{jj}$, and due to reciprocity $Z_{ij} = Z_{ji}$ for ($i$ and $j = 1,2,...,4$).

The modified open source voltages $V'_{oj}$ for ($j = 1,2,...,4$) in (5.14) of the MFSAA are calculated for the mutual coupling of both methods. Results from both active input impedance and conventional method for the mutual coupling and modified feed network of the MFSAA at 2.45 GHz are summarized in Tab.5.1. From the conventional method, all array elements have the same input impedance, while from the active input impedance method, the array elements input impedances are different depending on the amplitudes and phases of the excitation vector.

The mutual coupling at 2.45 GHz is nearly the same from both methods except for change in sign of the reactance of $Z_{12}$ (mirror effect as seen in Fig.5.1). The modified radiation patterns based on the feed networks calculated in Tab.5.1 from both methods at 2.45 GHz are shown in Fig.5.2 and compared with the coupled pattern of the excitation vector defined by (5.13).

For the radiation pattern of the excitation vector calculated from the conventional method, the nulls are filled, the side lobes level becomes higher, and the front to back ratio becomes lower compared to the coupled pattern. While for the radiation pattern of the excitation vector calculated from the active input impedance, the depth of the nulls is -27.15 dB compared to -13.8 dB for the coupled pattern, and the side lobes level is lower. The front to back ratio is -14.83 dB compared to -24 dB for the coupled pattern due to mismatched ports.

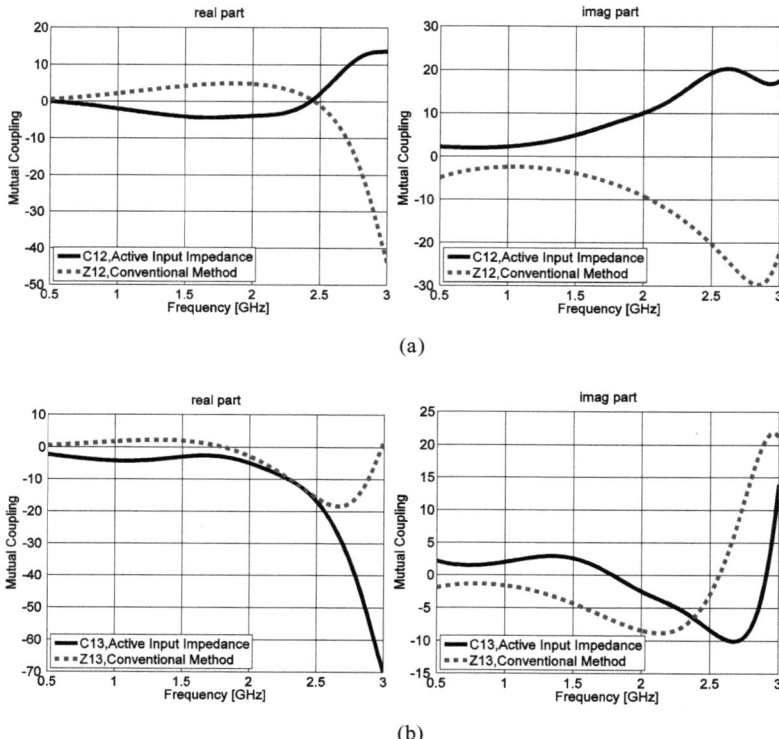

Figure 5.1: Mutual coupling of the MFSAA calculated from active input impedance (solid) and from conventional method (dash). (a) $Z_{12}, C_{12}$ between adjacent array elements, and (b) $Z_{13}, C_{13}$ between nonadjacent array elements.

The feed network obtained from the active input impedance is more effective in compensating for beamforming compared to the conventional method as Fig.5.2 shows. The MFSAA ports are still coupled. In the next section of this chapter, a design of decoupling and matching networks (DMN) of the MFSAA is proposed for the purpose of port isolation and their effect on the array patterns is discussed.

## 5.2 Compensating for Coupled Radiation Patterns of the MFSAA

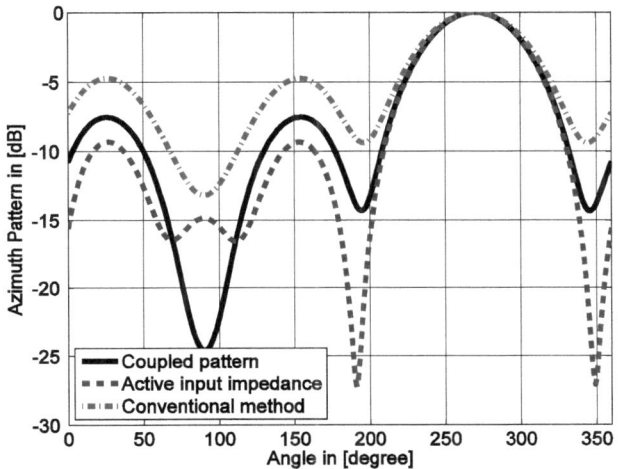

Figure 5.2: Azimuth pattern of the MFSAA coupled pattern (solid), active input impedance (dash), and conventional method (dash dot).

TABLE 5.1: MUTUAL COUPLING AND MODIFIED FEED NETWORK OF THE MFSAA AT 2.45 GHz CALCULATED FROM ACTIVE INPUT IMPEDANCE AND FROM CONVENTIONAL METHOD.

|  | Input Impedance | Self- and Mutual Impedances | Feed Network |
|---|---|---|---|
| Conv. Method | $Z_{in} = 52.225 + j7.81$ | $Z_{11} = 41.5 + j8.67$<br>$Z_{12} = 0.191 - j18.837$<br>$Z_{13} = -14.483 - j4.627$<br>$Z_{14} = Z_{12}$ | $V'_{o1} = 1.51e^{0.381°}$<br>$V'_{o2} = 0.856e^{87.709°}$<br>$V'_{o3} = 0.781e^{-175.709°}$<br>$V'_{o4} = V'_{o2}$ |
| Active In. Imp. Method | $Z_{active1} = 17.528 + j7.646$<br>$Z_{active2} = 47.309 + j3.651$<br>$Z_{active3} = 209.071 + j48.45$<br>$Z_{active4} = Z_{active2}$ | $Z_{11} = 41.5 + j8.67$<br>$C_{12} = 0.058 + j20.192$<br>$C_{13} = -15.725 - j5.079$<br>$C_{14} = C_{12}$ | $V'_{o1} = 0.648e^{2.185°}$<br>$V'_{o2} = 1.059e^{86.42°}$<br>$V'_{o3} = 1.653e^{-178.91°}$<br>$V'_{o4} = V'_{o2}$ |

## 5.3 Compensating for Port Coupling of the MFSAA

Port isolation is critical in many practical applications such as diversity systems and multiple input multiple output (MIMO) systems for the purpose of increase of the channel capacity in wireless communication systems. There are several methods that can be utilized to enhance port isolation. In [55, 56], the electromagnetic band gab structures (EBG) are used to suppress the surface wave between antenna elements and thus enhance the isolation between antenna ports. In [55], a mushroom-like EBG structure is inserted between a two-element microstrip antenna array for the purpose of surface waves suppression resulting in a low mutual coupling and better antenna array performance. The mushroom-like EBG structure consists of: a ground plane, a dielectric substrate, metallic patches, and connecting vias. However, the large area required for the EBG is a drawback for this technique of port isolation. In another research work [62-66], a decoupling and matching network (DMN) is designed based on the theory of eigenmode for good isolation between ports. While in [59, 67, 68], direct connection between antenna elements, and in [69-71], coupled elements are inserted between antenna elements to create an additional physical path in case of direct connection and a coupled path in case of coupled or parasitic elements for enhancing the isolation. Defected ground structures (DGS) are used in [57, 58, 71] to provide port isolation by suppressing the current flow between antenna elements which, however, makes the RF feed circuits more difficult. In [58], a defected ground structure was used to decouple a two-element microstrip antenna array. The DGS provides a wide band-stop characteristics that suppresses surface wave excited in microstrip antennas through the substrate layer (thick and high permittivity substrate).

### 5.3.1 Parasitic Elements Based Decoupling Technique

In [69], perfect decoupling between two arbitrary spaced antennas has been proposed using a reactively loaded parasitic antenna in between two active antennas. Tuning the reactive load of the parasitic element in addition to the dimensions of active and parasitic elements has provided perfect port isolation over a narrow frequency band. In [70], a coupling element with a total electrical length of a round half-wavelength at 2.4 GHz is added between two antenna elements. The coupling element dimensions are optimized to compensate for the induced current in the second element of the array due to the excited current in the array first element. A good isolation over a wider frequency band at 2.45 GHz is obtained compared to [69].

In [71], the concept of parasitic decoupling elements is applied to the MFSAA to obtain good port isolation over a wider frequency band. Instead of reactively loaded parasitic

## 5.3 Compensating for Port Coupling of the MFSAA

element [69], short circuited parasitic elements are proposed. By proper choice of parasitic elements positions and dimensions in addition to the array active elements dimensions, port isolation over a wider frequency range is obtained for a given array elements spacing without extending the overall area. The decoupled ports of the array can then be matched using conventional matching networks [23].

Consider an array of two active elements and one short-circuited parasitic element in between as in Fig.5.3. The terminal currents and voltages are related through the impedance matrix as

$$\begin{bmatrix} V_1 \\ V_2 \\ V_3 \end{bmatrix} = \begin{bmatrix} Z_{11} & Z_{12} & Z_{13} \\ Z_{21} & Z_{22} & Z_{23} \\ Z_{31} & Z_{32} & Z_{33} \end{bmatrix} \begin{bmatrix} I_1 \\ I_2 \\ I_3 \end{bmatrix} \quad (5.16)$$

The short-circuit condition for the parasitic element implies that $V_2 = 0$, and due to reciprocity, $Z_{ij} = Z_{ji}$, where $i \neq j$, $\{i, j = 1,2,3\}$. Substituting into (5.16) and rearrangement, the voltages and currents of array active elements are related as

$$\begin{bmatrix} V_1 \\ V_3 \end{bmatrix} = \begin{bmatrix} Z'_{11} & Z'_{13} \\ Z'_{13} & Z'_{33} \end{bmatrix} \begin{bmatrix} I_1 \\ I_3 \end{bmatrix}, \quad (5.17)$$

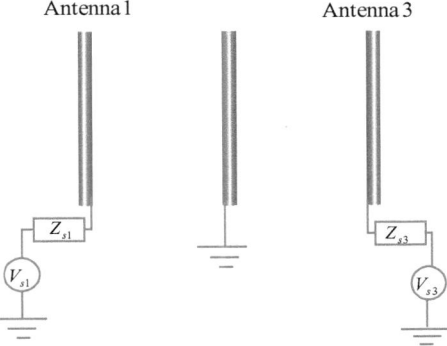

Figure 5.3: Two active elements array with one short parasitic element as a short-circuited decoupling element.

where for identical array elements

$$Z'_{11} = Z_{11} - \frac{Z_{12}^2}{Z_{22}} = Z'_{33}$$
$$Z'_{13} = Z_{13} - \frac{Z_{12}^2}{Z_{22}}.$$
(5.18)

For decoupled ports, the condition $Z'_{13} = 0$ should be satisfied which implies that $Z_{13}$ should equal to $Z_{12}^2/Z_{22}$. The array active and parasitic elements in Fig.5.3 are quarter wavelength monopoles, so tuning the parasitic element length and diameter can be used to compensate for the mutual coupling between array active elements. The input impedance of array decoupled ports is now mismatched to the source impedance ($50\,\Omega$) by the same factor $Z_{12}^2/Z_{22}$. The decoupled ports of the array can now be matched by using conventional matching circuits.

For the MFSAA, four parasitic elements are required to decouple adjacent array active elements and one parasitic element in the center to decouple nonadjacent array active elements as Fig.5.4 shows.

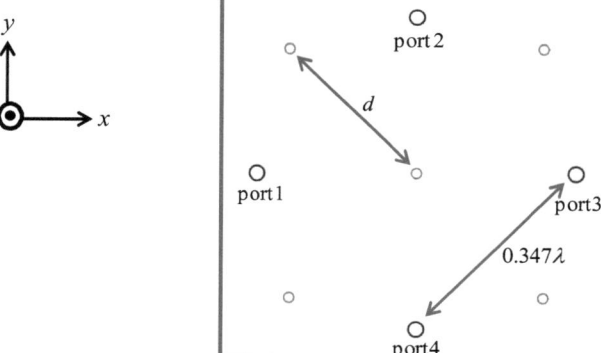

Figure 5.4: MFSAA with five short parasitic elements (pink) to decouple array active elements (black). The monopoles are quarter-wavelength at 2.45 GHz.

## 5.3 Compensating for Port Coupling of the MFSAA

For nonadjacent array active elements, the length and diameter of the parasitic element in the center of the array are optimized to compensate for mutual coupling between nonadjacent array active elements. While for adjacent array active elements, the length, diameter, and position of parasitic elements at distance $d$ from the center are optimized to get isolated ports. The positions of the four parasitic elements are optimized by adjusting the distance $d$ from the center which is the radius of the circle where the four parasitic elements are placed. Due to the symmetry of the MFSAA, the four parasitic elements placed between adjacent array elements at distance $d$ from the center should have the same length. The length and position of parasitic elements placed between adjacent array active elements, the length of the parasitic element at the center for nonadjacent array active elements, the diameter of the array parasitic elements in addition to the array active elements length for given separation distance of $0.347\lambda$ between adjacent array active elements are all optimized with Empire XCcel. The optimization goal is defined such that good isolation between antenna ports at 2.45 GHz is obtained. Port isolation of more than 30 dB as shown in Fig.5.5 is obtained for array active elements length of $0.2\lambda$, parasitic elements lengths of $0.253\lambda$ and $0.286\lambda$ between adjacent and nonadjacent array active elements respectively, and $d = 0.22\lambda$. The diameters of array active and parasitic elements are chosen to be $0.025\lambda$ and $0.016\lambda$ respectively. However, the decoupled ports are mismatched to the 50 $\Omega$ reference as seen in Fig.5.5.

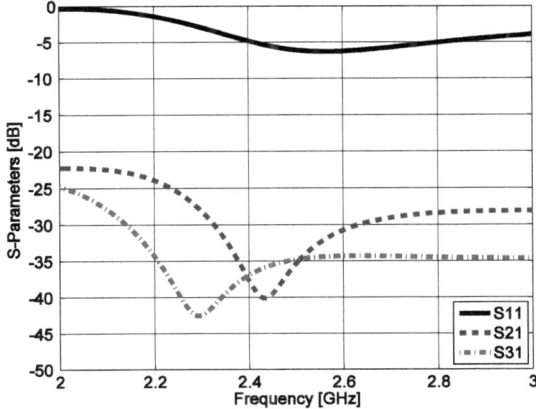

Figure 5.5: Simulated S-parameters of the MFSAA with parasitic decoupling elements.

A matching network of an open circuit stub is thus designed by optimizing the stub length and position from the feed point for good matching at 2.45GHz. For comparison, Fig.5.6 and Fig.5.7 show the MFSAA scattering parameters with and without parasitic decoupling elements and matching circuit.

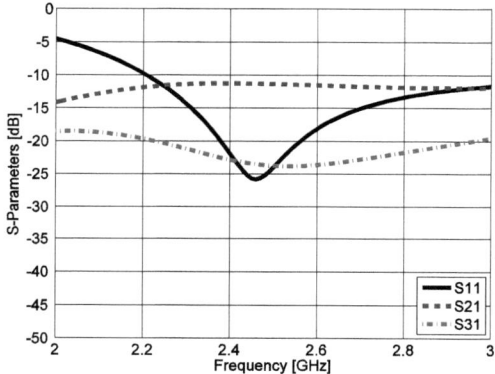

Figure 5.6: Simulated S-parameters of the MFSAA without parasitic decoupling elements and matching circuit.

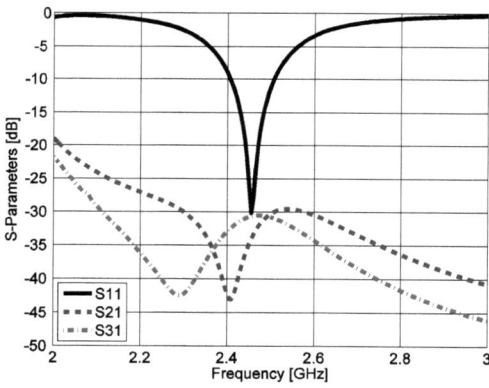

Figure 5.7: Simulated S-parameters of the MFSAA with parasitic decoupling elements and matching circuit.

## 5.3 Compensating for Port Coupling of the MFSAA

An isolation of more than 30 dB is obtained over a frequency band from 2.406 to 2.506 GHz (10 dB bandwidth). The scattering parameters ($S_{12}$ and $S_{13}$) represent the coupling between adjacent and nonadjacent array active elements respectively.

Azimuth and elevation patterns of port1 excited and all other ports matched to 50 Ω with and without parasitic decoupling elements are shown in Fig.5.8. Due to the additional parasitic elements, the element radiation pattern of the MFSAA with decoupled ports is more directive by 10 dB compared to the MFSAA element pattern with coupled ports. In the elevation plane, the null at $\theta = 0°$ is seen to be filled due to strong currents excited on the ground plane between the active and the parasitic elements which radiate vertically. Radiation contributions of the parasitic elements combine with the active element radiation to create a dip at $\theta = 45°$ for $\text{phi} = 0°$ plane.

The parasitic elements provide additional coupling paths to compensate for the mutual coupling effect between array elements. As port1 is excited and all other ports are matched terminated, an 180° out of phase current is induced in both the array elements and in the parasitic elements. A current in-phase to the current in the driven element (port1) is induced in array elements due to the current in the parasitic elements which cancels the 180° out of phase current from direct coupling to the driven element and thus achieves port isolation. The 10 dB higher directivity seen in Fig.5.8 of the decoupled MFSAA elements is due to the current in the short-circuited parasitic elements. Approximately equal magnitude and anti-phase currents in the driven element and parasitic element result in a higher directivity as known from the theory of superdirective arrays [73, 74]. The parasitic element provides a new resonant frequency $f_{par}$ and two superdirective frequencies emerge on either side of $f_{par}$ at which the parasitic element acts as a director or a reflector. Usually the parasitic superdirectivity is optimized by using the parasitic element as a reflector [75]. The optimization of the MFSAA with parasitic decoupling elements gives good port isolation over a wide frequency band for the parasitic elements as reflectors rather than directors. The new resonant frequency $f_{par}$ and the additional two superdirective frequencies explain the good port isolation over a wider frequency band as Fig.5.7 shows. Good port isolation due to the parasitic decoupling elements can also be seen from the electric field distribution in the xy-plane of the MFSAA at 2.45 GHz, shown in Fig.5.9, where the field close to the radiator elements is very low.

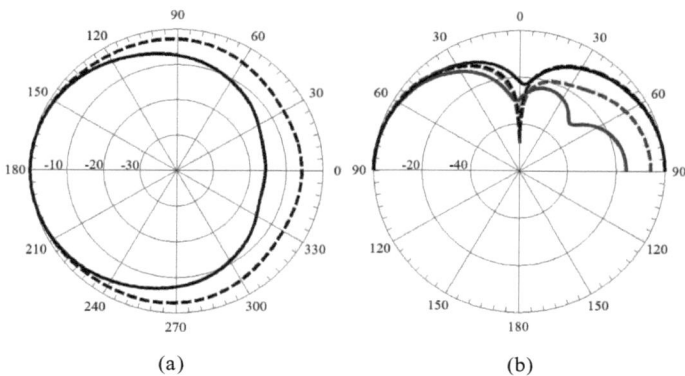

Figure 5.8: MFSAA radiation pattern with (solid) and without (dash) parasitic decoupling elements and matching circuit. (a) Azimuth pattern and (b) elevation pattern at $phi = 0°$ (blue) and $phi = 90°$ (black).

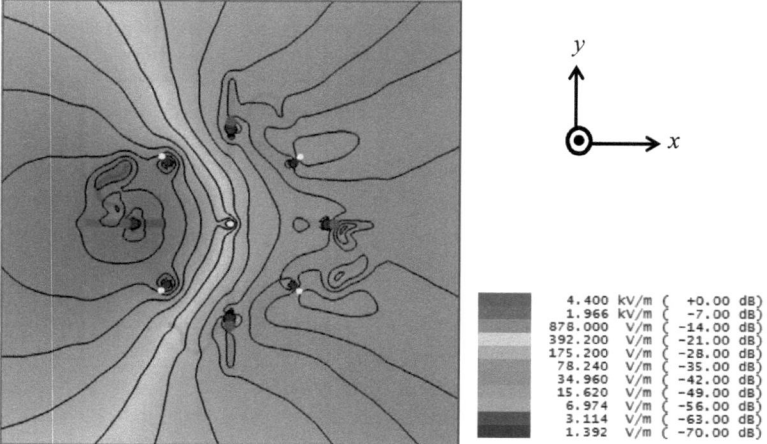

Figure 5.9: Electric field distribution ($E_{xyz}$) of the MFSAA with parasitic decoupling elements and matching circuit at 2.45 GHz.

## 5.3 Compensating for Port Coupling of the MFSAA

The simulated total efficiency is 94% for the MFSAA with parasitic decoupling elements and matching circuit compared to 84% for the MFSAA without parasitic decoupling elements and matching circuit. For MIMO systems, suppression of mutual coupling between array elements of the MFSAA using parasitic elements realizes in addition to the improved power transfer from the antenna elements to the loads an independent channel of operation with isolation of more than 30 dB between array ports of $0.347\lambda$ separation. An enhancement of more than 10 dB is achieved compared to conventional MIMO systems which show an isolation of about 20 dB for half wavelength separation between array elements [76, 77].

In above discussion, decoupling of the MFSAA using parasitic elements is discussed without considering the effect of chassis modes. For the MFSAA setting in Fig.5.4, exciting port1 while keeping all other ports matched couples to the half- and full-wavelength modes of the chassis end. The effect of those two chassis modes on the radiation pattern of the MFSAA can be seen from Fig.5.10. The pattern in the H-plane of the half-wavelength chassis mode presented in Fig.4.10a fills the dip at $\theta = 0°$ of the MFSAA elevation pattern while the dip at $\theta = 45°$ is filled due to the H-plane patterns of the half- and full-wavelength chassis modes presented in Fig.4.10a and b, respectively. Distortion in azimuth pattern of the MFSAA seen in Fig.5.10 is due to the E-plane patterns of the half- and full-wavelength chassis modes presented in Fig.4.10a and b, respectively.

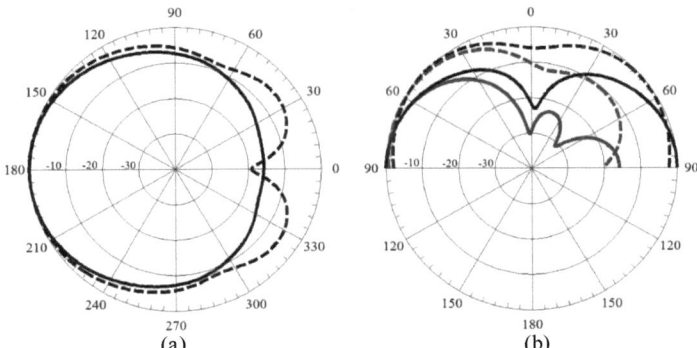

Figure 5.10: MFSAA radiation pattern with parasitic decoupling elements on infinite (solid) and finite (dash) ground plane. (a) Azimuth pattern and (b) elevation pattern at $\mathrm{phi}=0°$ (blue) and $\mathrm{phi}=90°$ (black).

### 5.3.2 Eigenmode Based Decoupling Network

The idea of connecting a lossless network between input ports and antenna ports such that there is no coupling between array elements was early proposed by [78]. For such a network to be realized, all the mutual antenna impedances should be reactive. Later on, different lossless decoupling networks based on the eigenmode theory were proposed in [62, 65] for antenna arrays with pure imaginary mutual impedances. For this concept, the array elements geometry and the distance between array elements have to be adjusted such that the array mutual impedances are pure imaginary. An array of $N$ elements produces $N$ distinct eigenmode. As the number of array elements increases, the array eigenmode also increases and it becomes difficult to get reactive mutual impedances. In [66], an eigenmode based decoupling network for the MFSAA was presented for arbitrary array mutual impedances. Optimization by ADS was performed for good port isolation. With the results obtained by the eigenmode theory represent the initial values for the optimization process. The goals are defined such that minimum coupling (less than -30 dB) at 2.45 GHz between adjacent and nonadjacent array elements is obtained. In the following, the method is presented in detail. The MFSAA admittance $Y_{array}$ matrix is given by

$$Y_{array} = \begin{bmatrix} Y_{11} & Y_{12} & Y_{13} & Y_{14} \\ Y_{21} & Y_{22} & Y_{23} & Y_{24} \\ Y_{31} & Y_{32} & Y_{33} & Y_{34} \\ Y_{41} & Y_{42} & Y_{43} & Y_{44} \end{bmatrix}. \tag{5.19}$$

For the MFSAA shown in Fig.4.1, due to reciprocity, $Y_{ij} = Y_{ji}$, where $i \neq j$, $\{i, j = 1,2,3,4\}$. The array is symmetric and the array elements are identical. The array admittance matrix $Y_{array}$ can be diagonalized as

$$Y_{array} = U Y_m U^T, \tag{5.20}$$

where $U$ is the orthonormal eigenvectors matrix given by

$$U = \frac{1}{2} \begin{bmatrix} 1 & 1 & 1 & 1 \\ 1 & 1 & -1 & -1 \\ 1 & -1 & -1 & 1 \\ 1 & -1 & 1 & -1 \end{bmatrix}. \tag{5.21}$$

## 5.3 Compensating for Port Coupling of the MFSAA

$Y_m$ is the modal admittance matrix of the MFSAA

$$Y_m = \begin{bmatrix} Y_{m1} & 0 & 0 & 0 \\ 0 & Y_{m2} & 0 & 0 \\ 0 & 0 & Y_{m3} & 0 \\ 0 & 0 & 0 & Y_{m4} \end{bmatrix}, \quad (5.22)$$

where $Y_{mi}$, $i = (i = 1,2,3,4)$ represents the $i\,th$ eigenvalue (modal admittance) of the MFSAA. Solving for the array modal admittances in (5.20) we have that

$$\begin{aligned} Y_{m1} &= Y_{11} + 2Y_{12} + Y_{13} \\ Y_{m2} &= Y_{11} - Y_{13} = Y_{m3} \\ Y_{m4} &= Y_{11} - 2Y_{12} + Y_{13}. \end{aligned} \quad (5.23)$$

Numerical simulations for frequency dependent mode admittances of the MFSAA structure in Fig.4.1 were carried out with Empire XCcel, as Fig.5.11 shows.

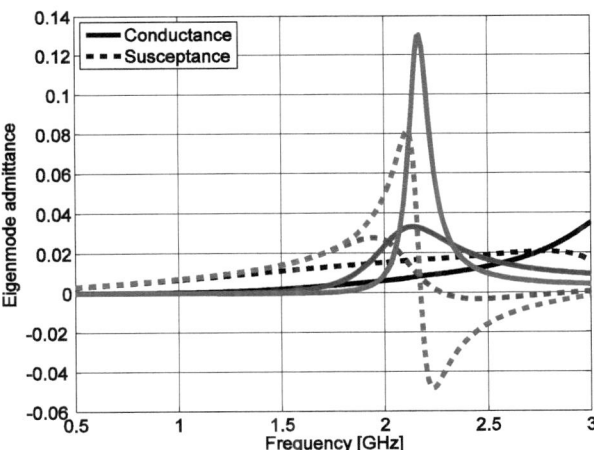

Figure 5.11: Frequency dependent MFSAA eigenmode admittances ($\Omega^{-1}$) $Y_{m1}$ (black), $Y_{m2,3}$ (blue), and $Y_{m4}$ (green).

Due to mutual coupling between array elements, the modal admittances are different from each other, while the load admittances are the same. In this case a complex mismatch modal factor $\Lambda_{mi}$ for $i = (1, 2, 3, 4)$ can be defined as

$$|\Lambda_{mi}|^2 = 1 - |\Gamma_{mi}|^2, \qquad (5.24)$$

where the corresponding modal reflection coefficient $\Gamma_{mi}$ is given by

$$\Gamma_{mi} = \frac{Z_L - Z_{mi}}{Z_L + Z_{mi}}. \qquad (5.25)$$

Substituting (5.25) in (5.24) implies that

$$\Lambda_{mi} = \frac{2\sqrt{R_L R_{mi}}}{Z_L + Z_{mi}} \qquad (5.26)$$

The radiation pattern for port $n$ is a linear combination of the mutually orthogonal eigen radiation patterns of the MFSAA where $n = (1, 2, 3, 4)$ [79]:

$$C_n(\theta, \phi) = \sum_{i=1}^{N} u_{ni} \Lambda_{mi} C_{mi}(\theta, \phi), \qquad (5.27)$$

where $u_{ni}$ are the elements of the unitary matrix $U$ defined in (5.21) and $C_{mi}(\theta, \phi)$ are the corresponding mutually orthogonal eigen radiation patterns of the MFSAA. $N$ is the number of distinct eigenmodes and $N = 4$ for the MFSAA. The decoupling of the MFSAA based on the eigenmode theory requires matching all modal admittances to the same admittance $(\Lambda_{mi} = 1)$ for $i = (1, 2, 3, 4)$. In [65], an $N$-element array with $k$ distinct eigenvalues required a decoupling network in $(k-1)$ stages where each stage consists of $2(k-1)$ series reactive elements $X$ and $2(k-1)$ shunt reactive elements $B$. For the MFSAA, two stages of a decoupling network are required: In the 1st stage modes (1, 4) admittances are matched, while in the 2nd stage modes (1, 4) and modes (2, 3) are matched. The 1st and 2nd stages of the decoupling network for the MFSAA are shown in Fig.5.12a and b respectively. From the 1st stage, the modal admittances $Y'_{mi}$, $i = (i = 1,2,3,4)$ as seen from the input ports $(1', 2', 3', 4')$ are obtained

$$Y'_{mi} = \left[Y_{mi}^{-1} + jX'_i\right]^{-1} + jn'_i B_1, \qquad (5.28)$$

## 5.3 Compensating for Port Coupling of the MFSAA

where $n'_i$ is an integer given by

$$n'_i = \begin{cases} 0 & \text{for mode 1} \\ 1 & \text{for mode 2,3} \\ 2 & \text{for mode 4.} \end{cases} \quad \text{for } (i = 1,2,3,4) \tag{5.29}$$

The value of $n'_i$ for $(i = 1,2,3,4)$ in (5.29) is calculated based on the corresponding eigenvector in (5.21). From the 2$^{nd}$ stage as Fig.5.12b shows, the modal admittances $Y''_{mi}$, $i = (i = 1,2,3,4)$ as seen from the input ports ( 1",2",3",4" ) are obtained as

$$Y''_{mi} = \left[ Y'^{-1}_{mi} + jX_2 \right]^{-1} + jn''_i B_2, \tag{5.30}$$

where $n''_i$ is an integer given by

$$n''_i = \begin{cases} 0 & \text{for mode 1} \\ 1 & \text{for mode 2,3} \\ 1 & \text{for mode 4.} \end{cases} \quad \text{for } (i = 1,2,3,4) \tag{5.31}$$

The value of $n''_i$ for $(i = 1,2,3,4)$ in (5.31) is calculated based on the corresponding eigenvector in (5.21). Closed-form design equations for calculating the values of reactive elements $X$ and $B$ in each stage are derived in [64] as

$$X = \frac{-b \pm \sqrt{b^2 - 4ac}}{2a} \tag{5.32}$$

$$B = \frac{1}{n} \left( \frac{X_{m2} + X}{R^2_{m2} + (X_{m2} + X)^2} - \frac{X_{m1} + X}{R^2_{m1} + (X_{m1} + X)^2} \right), \tag{5.33}$$

where

$$\begin{aligned} n &= n_1 - n_2 \\ a &= R_{m1} - R_{m2} \\ b &= 2(R_{m1} X_{m2} - R_{m2} X_{m1}) \\ c &= R_{m1}(R^2_{m2} + X^2_{m2}) - R_{m2}(R^2_{m1} + X^2_{m1}). \end{aligned} \tag{5.34}$$

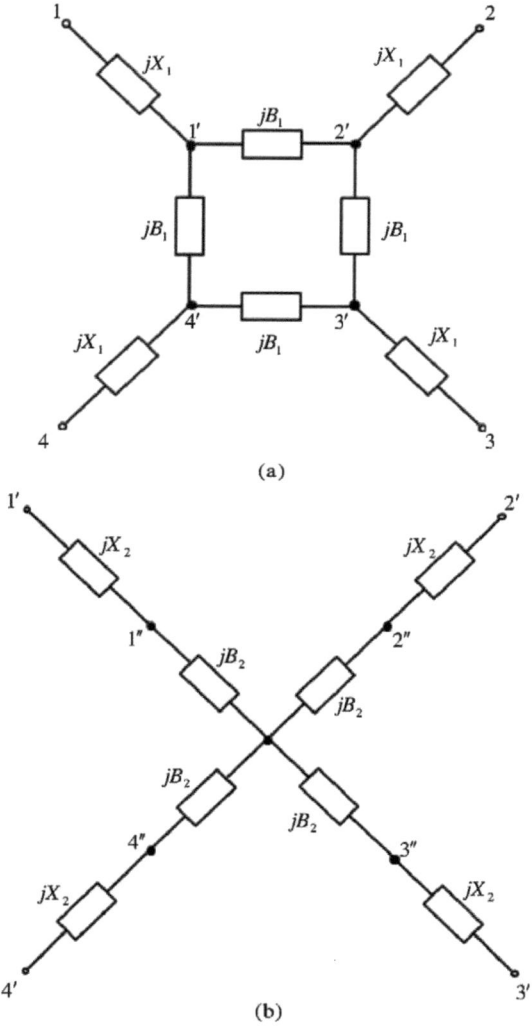

Figure 5.12: Decoupling network of the MFSAA. (a) 1$^{st}$ stage and (b) 2$^{nd}$ stage.

## 5.3 Compensating for Port Coupling of the MFSAA

All modes are now matched to the same input admittance as seen from the input ports ( $1'',2'',3'',4''$ ). External matching circuits (L-section matching circuit) are required to match all modal impedances to the source impedance ($50\,\Omega$). Impedance parameters and reactive elements values of the decoupling and matching network for the MFSAA are summarized in Tab.5.2. The admittance matrices $Y'_D$ and $Y''_D$ for the 1$^{st}$ and 2$^{nd}$ stages respectively of the decoupling network are obtained to be

$$Y'_D = \begin{bmatrix} jB_1 & -j\frac{1}{2}B_1 & 0 & -j\frac{1}{2}B_1 \\ -j\frac{1}{2}B_1 & jB_1 & -j\frac{1}{2}B_1 & 0 \\ 0 & -j\frac{1}{2}B_1 & jB_1 & -j\frac{1}{2}B_1 \\ -j\frac{1}{2}B_1 & 0 & -j\frac{1}{2}B_1 & jB_1 \end{bmatrix} \quad (5.35)$$

$$Y''_D = \begin{bmatrix} j\frac{3}{4}B_2 & -j\frac{1}{4}B_2 & -j\frac{1}{4}B_2 & -j\frac{1}{4}B_2 \\ -j\frac{1}{4}B_2 & j\frac{3}{4}B_2 & -j\frac{1}{4}B_2 & -j\frac{1}{4}B_2 \\ -j\frac{1}{4}B_2 & -j\frac{1}{4}B_2 & j\frac{3}{4}B_2 & -j\frac{1}{4}B_2 \\ -j\frac{1}{4}B_2 & -j\frac{1}{4}B_2 & -j\frac{1}{4}B_2 & j\frac{3}{4}B_2 \end{bmatrix} \quad (5.36)$$

TABLE 5.2: MFSAA IMPEDACE PARAMETERS AND ITS DECOUPLING AND MATCHING NETWORK REACTIVE ELEMENT VALUES.

| MFSAA input impedance parameters at matching frequency 2.45GHz | $Z_{11} = 46.867 + j18.502$ <br> $Z_{12} = 2.603 - j23.268$ <br> $Z_{13} = -17.268 - j9.061$ |
|---|---|
| Decoupling network 1$^{st}$ stage | $X = -11.941, \quad C = 5.440\,pF$ <br> $B = 0.016, \quad C = 1.005\,F$ |
| Decoupling network 2$^{nd}$ stage | $X = 34.173, \quad L = 2.220\,nH$ <br> $B = 0.011, \quad C = 0.711\,pF$ |
| L-section matching circuit | $X = 51.191, \quad L = 2.298\,nH$ <br> $B = 0.021, \quad C = .868\,pF$ |

Capacitance and inductance values of the MFSAA decoupling matching network are calculated based on the values of reactive elements $X$ and $B$ in Tab.5.2. The circuit model in ADS for the eigenmode based decoupling and matching network of the MFSAA is presented in Appendix D. The scattering parameters of the MFSAA with and without a decoupling matching network are shown in Fig. 5.13a and b, respectively. Port isolation of more than 20 dB is obtained over a frequency range from 2.37 to 2.53 GHz.

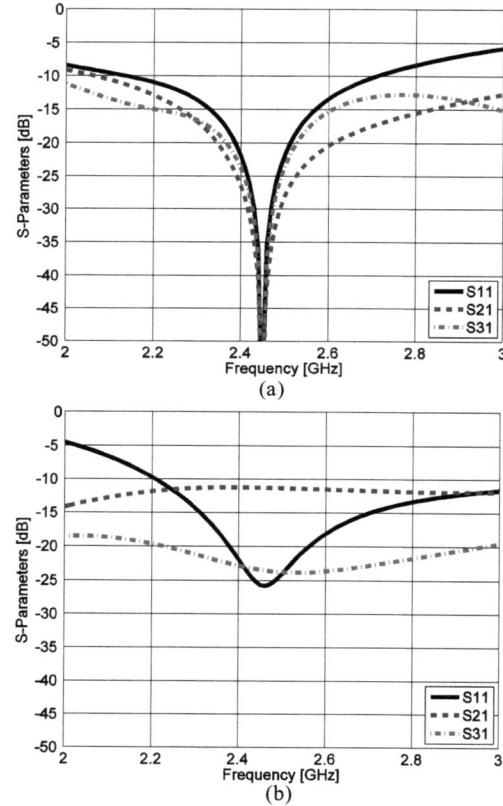

Figure 5.13: S-parameters of the MFSAA. (a) With and (b) without a decoupling matching network.

## 5.3 Compensating for Port Coupling of the MFSAA

The decoupling and matching network designed above was based on lossless concentrated capacitors and inductors. This network of ideal lumped elements has to be converted to a physically realizable form. One convenient way of realization uses microstrip line (MSL) for connecting ports and realizing reactive elements by replacing the capacitors and inductors with open and short circuit stubs calculated so that

$$\theta_{sc} = \tan^{-1}(\frac{\omega L}{Z_o})$$
$$\theta_{oc} = \tan^{-1}(\omega C Z_o). \tag{5.37}$$

A microstrip line decoupling matching network was designed and optimized so that good isolation between ports was achieved. Fig. 5.14 shows the scattering parameters of the MFSAA decoupling matching network realized in MSL. The layout of the MSL realization is shown in Appendix E. Port isolation of more than 20 dB is obtained over a narrower frequency band from 2.427 GHz to 2.453 GHz because of the frequency dependent transmission line transformation of the microstrip line compared to the more static transformation of the lumped elements decoupling matching network.

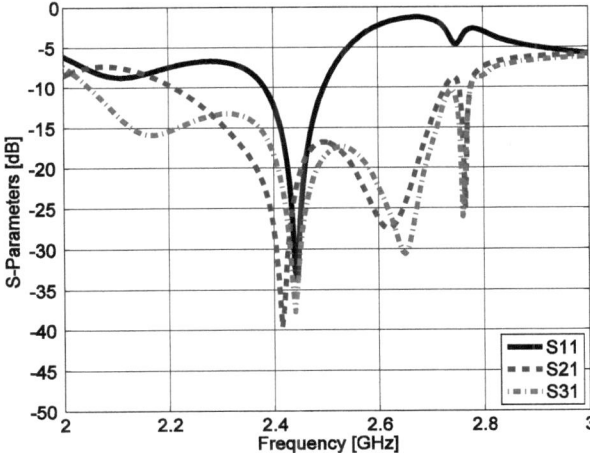

Figure 5.14: S-parameters of the MFSAA with the decoupling and matching network of Fig. 5.12 realized in MSL.

Azimuth and elevation patterns of port1 excited and all other ports matched to 50 Ω with and without an eigenmode based decoupling and matching network are shown in Fig.5.15. The element radiation pattern of the MFSAA with decoupled ports is more directive by approximately 10 dB compared to the MFSAA element pattern with coupled ports. In the eigenmode based decoupling network, a direct connection between array elements is modelled to create a physical path compared to the coupled path in the case of the parasitic decoupling elements presented in section 5.3.1 for port isolation. When port1 is excited and all other ports are matched to 50 Ω, 180° out of phase currents are induced in the other array elements compensating the induced currents and resulting in good port isolation.

For the parasitic decoupling elements, isolation over wider frequency band is obtained compared to the eigenmode based decoupling network realized in MSL. Since the microstrip line realization limits the freedom of optimizing for a wider frequency band. In [80], each reactive element in the eigenmode based decoupling network is replaced by a series or parallel combination of an inductor and a capacitor to match over a wider frequency band.

Good port isolation with the eigenmode based decoupling network can also be seen from the electric field distribution in xy-plane of the MFSAA at 2.45 GHz as shown in Fig.5.16. The simulated total efficiency is 96% for the MFSAA with an eigenmode based decoupling and matching network compared to 84% for the MFSAA without a decoupling and matching network.

The eigenmode based decoupling network of the MFSAA provides good port isolation at 2.45 GHz and maximum power transfer between the 50 Ω source and the antenna elements is obtained. The azimuth radiation pattern of the MFSAA with an eigenmode based decoupling and matching network and excitation vector of (4.4), for a beamforming in $\phi = 45°$ and nulls in $\phi = 135°, 225°,$ and $315°$ directions can be seen in Fig.5.17. For beamforming compared to the isolated elements azimuth pattern, Fig.4.5, the nulls are filled but the patterns are decoupled. As discussed in section 5.2, the excitation vector of the MFSAA has to be optimized to compensate for beamforming.

## 5.3 Compensating for Port Coupling of the MFSAA    91

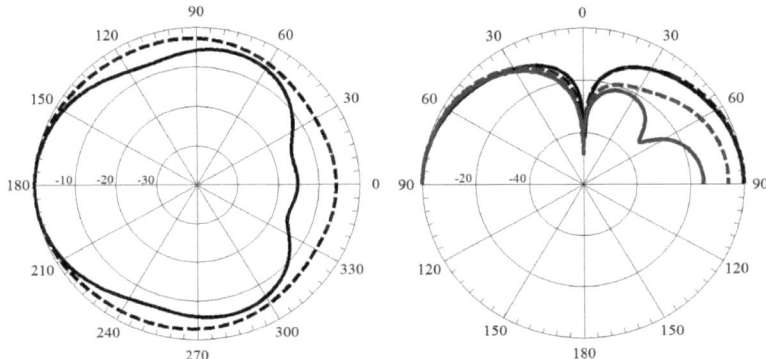

Figure 5.15: MFSAA radiation pattern with (solid) and without (dash) a MSL decoupling matching network. (a) Azimuth pattern and (b) elevation pattern at $\text{phi}=0°$ (blue) and $\text{phi}=90°$ (black).

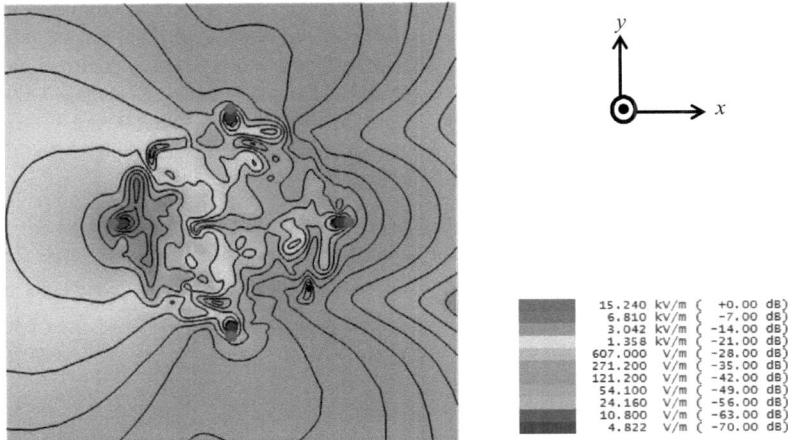

Figure 5.16: Electric field distribution ($E_{xyz}$) of the MFSAA with a MSL decoupling matching network at 2.45 GHz.

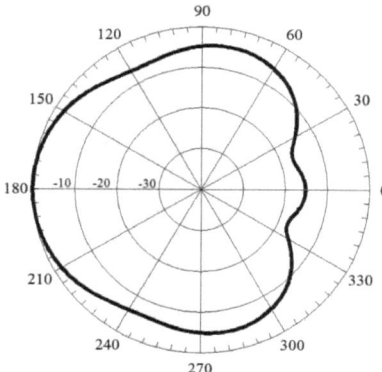

Figure 5.17: Azimuth radiation pattern of the MFSAA with an eigenmode based decoupling network and excitation vector of (4.4).

Mutual coupling between array elements reduces the signal to noise ratio of the system (SNR) especially when the spacing between antenna elements is less than half a wavelength [62]. Modification of the excitation vector in beamforming and null steering antenna arrays cannot compensate for SNR degradation as can be concluded from the MFSAA pattern with the excitation vector obtained from the active input impedance method seen in Fig.5.2 compared to the isolated azimuth pattern of the MFSAA shown in Fig.4.5. Therefore, implementation of a decoupling and matching circuit between the input ports and antenna ports is essential for high SNR.

The excitation vector of the MFSAA with an eigenmode based decoupling and matching circuit is optimized by the help of EM simulation (Empire Xccel) to compensate for modified patterns of the MFSAA for the purpose of beamforming in $\phi = 45°$ and nulls in $\phi = 135°, 225°,$ and $315°$ directions. The optimized excitation vector obtained from EM simulation for isolated ports MFSAA is given by

$$\begin{bmatrix} V_{o1} \\ V_{o2} \\ V_{o3} \\ V_{o4} \end{bmatrix} = \begin{bmatrix} 0.28 \\ 0.33 e^{j\pi/2} \\ e^{j1.72\pi/2} \\ 0.33 e^{j\pi/2} \end{bmatrix} \tag{5.38}$$

## 5.3 Compensating for Port Coupling of the MFSAA 93

The azimuth beam of the MFSAA corresponding to the excitation vector in (5.38) is shown in Fig.5.18. A null deep of less than -40 dB is obtained. Isolated ports and modified patterns of the MFSAA at 2.45 GHz without mutual coupling effects are now obtained as can be seen from Fig.5.14 and Fig.5.18 respectively. The three other beams are generated by cyclic rotation of the port excitation.

The MFSAA with an eigenmode based decoupling network was discussed without considering the effect of chassis modes. To consider the effect of chassis modes coupling, exciting port1 while keep all other ports matched terminated couples to the half- and full-wavelength modes of the chassis end. The effect of those two chassis modes on the radiation pattern of the MFSAA can be seen from Fig.5.19.

The H-plane pattern of the half-wavelength chassis mode presented in Fig.4.10a fills the null at $\theta = 0°$ of the MFSAA elevation pattern. The null at $\theta = 45°$ is filled due to the H-plane patterns of the half- and full-wavelength chassis modes presented in Fig.4.10a and b, respectively. Distortion in azimuth pattern of the MFSAA seen in Fig.5.19 is due to the E-plane patterns of the half- and full-wavelength chassis modes presented in Fig.4.10a and b, respectively.

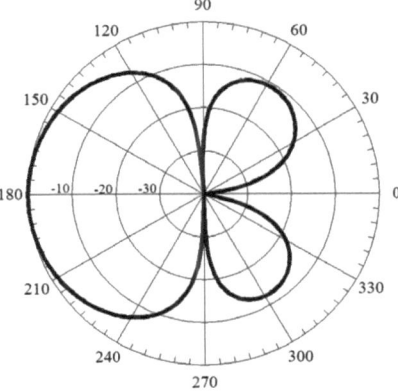

Figure 5.18: Azimuth beam of the MFSAA with an eigenmode based decoupling network and excitation vector of (5.38).

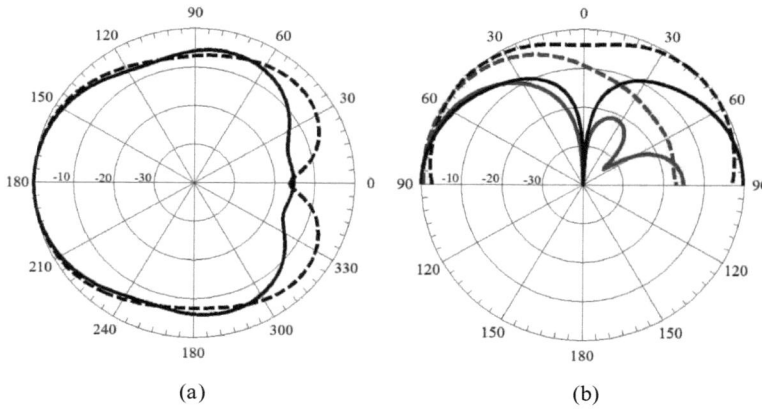

Figure 5.19: MFSAA radiation pattern with an eigenmode based decoupling network at infinite (solid) and finite (dash) ground plane. (a) Azimuth pattern and (b) elevation pattern at $phi = 0°$ (blue) and $phi = 90°$ (black).

# Chapter 6

# Utilization of the Characteristic Mode Theory for MIMO Antenna Systems

## 6.1 Introduction

As shown in previous chapters, the modification of the excitation vector can compensate for the effect of mutual coupling in the MFSAA for the purpose of beamforming to suppress the interfering signals and improve the performance of wireless communication systems. Mutual coupling between the array elements reduces the signal to noise ratio of the system (SNR) especially when the spacing between antenna elements is less than half a wavelength as in the case of handheld devices such as smartphones. Modification of the excitation vector in beamforming and null steering antenna arrays cannot compensate for SNR degradation. In such a situation, implementation of decoupling matching networks between the input ports and antenna ports is required for matching port impedances and compensating for reduction in the SNR. In chapter five, parasitic decoupling elements and eigenmode based decoupling networks are implemented and discussed for the MFSAA. In this chapter, an alternative approach to achieve port decoupling is proposed. It utilizes the theory of characteristic modes. Port isolation is obtained by making use of the orthogonality of characteristic wavemodes of the chassis.

In MIMO systems, the use of multiple antennas can result in an enhancement with respect to the channel capacity and data rate. Multiport antenna systems usually have the design goal of port isolation and pattern diversity. The demand of antenna design for MIMO goes towards the coupling-element based antenna structures, where the chassis is the main radiator mainly at low frequencies [14, 39, 40]. In coupling-element based antenna as discussed in chapter two, the coupling element is a small volume nonresonant antenna used to excite the chassis orthogonal current modes through coupling to the electric field (capacitive coupling) or to the magnetic field (inductive coupling) [21]. In [81], a two-port MIMO antenna is proposed with two capacitive coupling elements to excite the first two dominant characteristic modes of a 100 mm × 40 mm chassis. The CCEs shape and location have to be properly chosen for strong coupling to the $1^{st}$ and $2^{nd}$ chassis modes. By the help of the characteristic mode theory, the characteristic modes of a 100 mm × 40 mm chassis can be numerically evaluated [11] and the location of the CCEs on a chassis can be properly chosen for optimal coupling to the dominant chassis wavemodes [16, 17]. By making use of the orthogonality of chassis wavemodes, port isolation between array elements is obtained.

## 6.2 Geometry of a Two-Port MIMO Antenna

For the two-port MIMO antenna in [81], the characteristic mode theory was applied to analyze a wire grid model of a 100 mm × 40 mm conducting plate [11]. The half- and full-wavelength resonant frequencies of the chassis major axis were found as 1.26, and 2.68 GHz and their respective radiation quality factors values are 2.3, and 3.0. In order to selectively excite the half- and full-wavelength modes of the chassis major axis, suitable couplers arrangement on the chassis and their excitation phases should be optimized [20]. For the two-port MIMO antenna in Fig.6.1, two CCEs are placed at the middle of the chassis short ends for a selective excitation to the half- and full-wavelength modes of the chassis major axis.

In [19, 82], for a selective excitation of the characteristic modes of the chassis near to their resonant frequencies, an arrangement of two and four capacitive coupling elements and a modal feed network has been proposed. A 180° hybrid based feed network was used to feed the structure by the eigenvectors of the selected chassis modes excitation. The S-parameters of the two-port antenna system in Fig.6.1 given by

$$S = \begin{bmatrix} S_{11} & S_{12} \\ S_{12} & S_{11} \end{bmatrix} \qquad (6.1)$$

The off-diagonal elements of S-matrix represent the mutual coupling between ports. Due to the symmetry properties of the structure, the S-matrix can be decomposed into

## 6.2 Geometry of a Two-Port MIMO Antenna

$$S = U\Gamma U^T, \quad (6.2)$$

where $U$ is the orthonormal eigenvectors matrix given by

$$U = \frac{1}{\sqrt{2}}\begin{bmatrix} 1 & 1 \\ 1 & -1 \end{bmatrix}, \quad (6.3)$$

and the diagonal matrix $\Gamma$

$$\Gamma = \begin{bmatrix} S_{11} + S_{12} & 0 \\ 0 & S_{11} - S_{12} \end{bmatrix}. \quad (6.4)$$

The two-column vectors of $U$ represent the excitation vectors for the half- and full-wavelength characteristic modes of the chassis. The modal feed network for the two-port antenna system in Fig.6.1 may simply be modeled as a 180° hybrid with S-parameters given by

$$S^m = \frac{1}{\sqrt{2}} \begin{bmatrix} 0 & 0 & 1 & 1 \\ 0 & 0 & 1 & -1 \\ 1 & 1 & 0 & 0 \\ 1 & -1 & 0 & 0 \end{bmatrix} \quad (6.5)$$

The CCEs are off-ground elements with a ground clearance of 10 mm. The off-ground antenna element provides better matching to the 50 Ω impedance at the design frequency of 1.8 GHz than the on-ground element [25, 27]. The dimensions of the whole structure are 120×40×5 mm$^3$. The CCE is bent from the top for better coupling to the chassis modes. The 180° hybrid of the modal feed network is replaced by a branchline coupler with an extra phase shift of 90° between its output ports to excite in phase corresponding to the even chassis mode, and 180° out of phase corresponding to the odd chassis mode eigenvectors.

The S-parameters of the two-port antenna with and without a modal feed network can be seen from Fig.6.2. Good port isolation of more than 20dB between antenna ports at 1.8 GHz is obtained. The modal feed network presented here supports pattern diversity between the chassis modes radiation patterns. The same concept could be applied for pattern diversity between array eigenmode radiation patterns for infinite ground [83].

## 98　Ch6: Utilization of the Characteristic Mode Theory for MIMO Antenna Systems

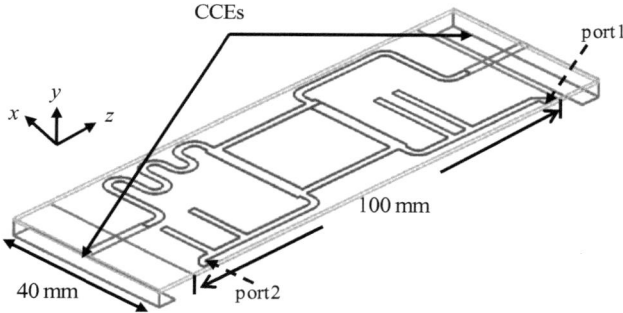

Figure 6.1: Geometry of a two-port center fed antenna structure with a double-stub matching circuit.

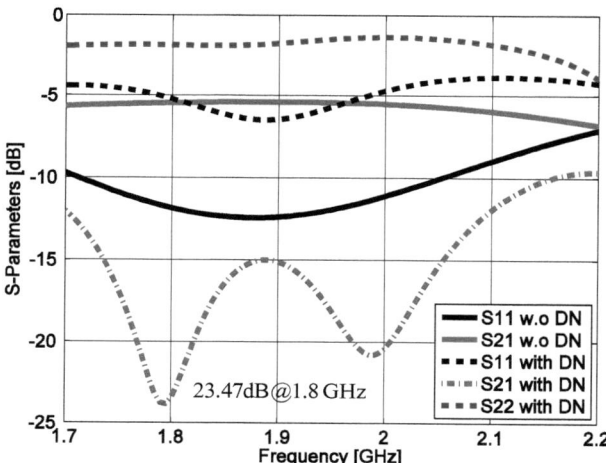

Figure 6.2: S-parameters of the two-port antenna in Fig.6.1 with and without a modal feed network.

## 6.3 Different Two-Port MIMO Antenna Configurations

The modal feed network discussed above provides decoupled antenna ports by feeding the structure with the first two orthogonal eigenvectors of the chassis. An external matching network is then required to match the two ports to the 50 $\Omega$. Matching over a wider frequency band depends on the number of lumped elements in the matching circuit, if an L-section matching circuit is chosen, then multiple L-sections may be required for sufficient frequency band. In Fig.2.12, the potential bandwidth for a single CCE placed at the middle of a 100mm×40mm chassis short end is plotted as a function of the number of lumped elements in the external matching circuit at 0.9 GHz and 2.2 GHz.

In [19], the authors aim to solve for wider impedance bandwidth by modifying the ground plane. Slots with optimized widths and lengths were created in the ground for better bandwidth. Modification in the ground plane will modify the modes current distributions of the chassis which in turns modify the corresponding radiation patterns of the chassis modes.

In [81], a comparison between two different configurations of a two-port MIMO antenna with respect to the port isolation and radiation patterns is presented. In the first configuration, the case of slotted ground plane with a single-stub as an external matching circuit is considered, while in the second configuration slots in the ground plane are replaced by a more complicated external matching circuit realized in microstrip line as a double-stub. A third configuration of a two-port MIMO antenna with an offset feed and a single-stub matching circuit is considered in this chapter and compared to the work presented in [81].

### 6.3.1 Two-Port MIMO Antenna with a Slotted Ground Plane

In this two-port antenna configuration, four slots with 1 mm width are made in the ground plane. Two slots of 0.05 $\lambda$ length are created at the middle of the chassis long ends, while two slots of 0.21$\lambda$ length are created at the middle of the chassis short ends. The CCEs are off-ground elements with a ground clearance of 13.5 mm. The dimensions of the whole structure are 127×40×6 mm$^3$. A single-stub matching circuit is designed to match the ports to the 50 $\Omega$ impedance. The top and bottom view of the two-port MIMO antenna with a slotted ground is shown in Fig.6.3. The single-stub position and length for each port, the ground clearance of the CCEs, the CCEs height, and the slots length are all optimized for good matching and port isolation at 1.925 GHz.

An isolation of more than 20 dB is obtained over a frequency band from 1.899 to 1.948 GHz (6 dB bandwidth), Fig.6.4.

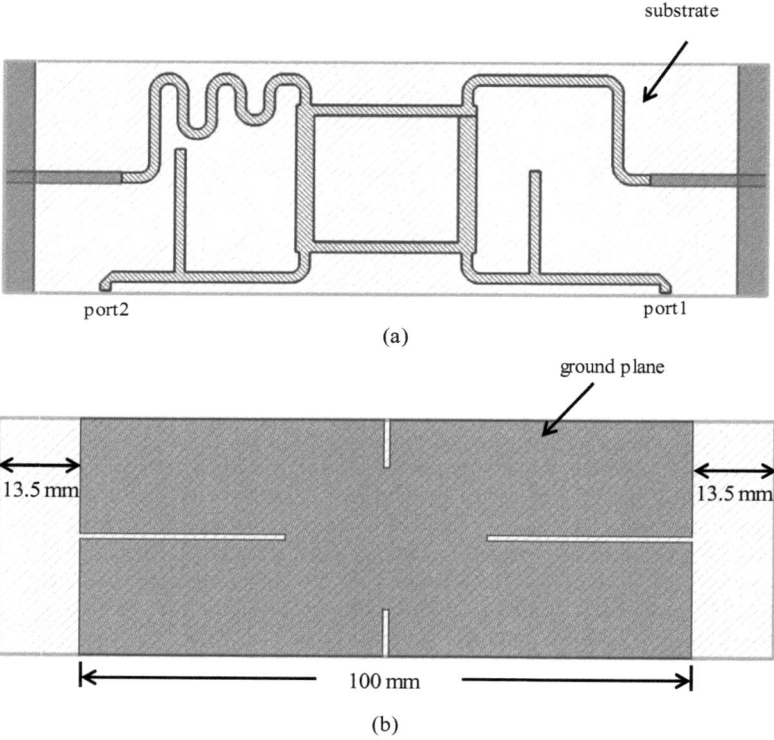

Figure 6.3: Top and bottom view of the two-port MIMO antenna with a slotted ground and a single-stub matching circuit.

## 6.3 Different Two-Port MIMO Antenna Configurations

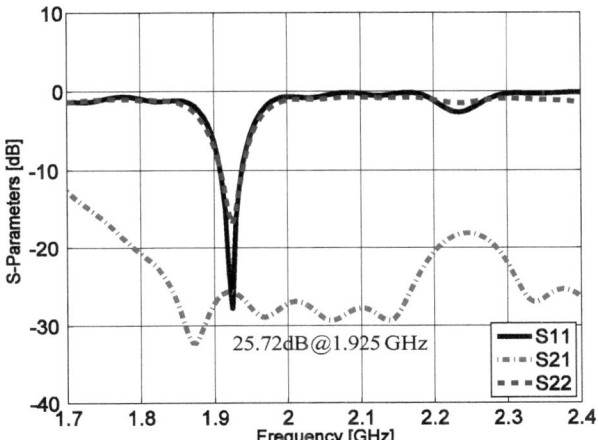

Figure 6.4: S-parameters of the two-port MIMO antenna with a slotted ground and a single-stub matching circuit.

### 6.3.2 Two-Port MIMO Antenna with a Double-Stub Matching Circuit

In this two-port MIMO antenna configuration the four slots in the slotted ground configuration discussed above are replaced by a double-stub external matching circuit to match the antenna ports to the 50 Ω. The CCEs are off-ground elements with a ground clearance of 10 mm. The dimensions of the whole structure are 120×40×4 mm$^3$. The top and bottom view of the two-port MIMO antenna with a double-stub matching circuit is shown in Fig.6.5. The double-stub position and length for each port, the ground clearance of the CCEs, and the CCEs height are all optimized for good matching and port isolation at 1.90 GHz. An isolation of more than 20 dB is obtained (6 dB bandwidth) over a frequency band from 1.887 to 1.925 GHz for the half-wavelength chassis mode and from 1.852 to 1.954 GHz for the full-wavelength chassis mode, Fig.6.6.

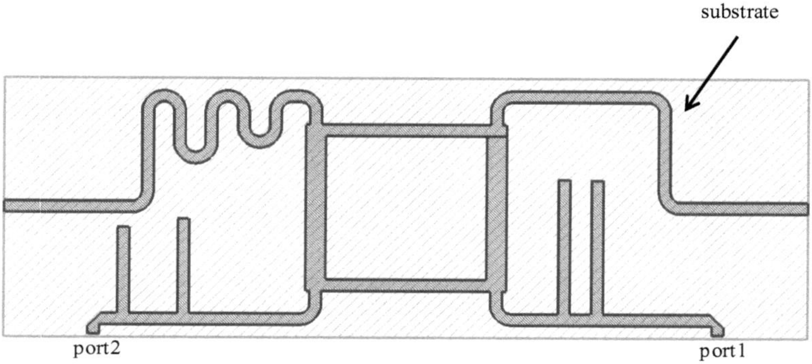


(a)

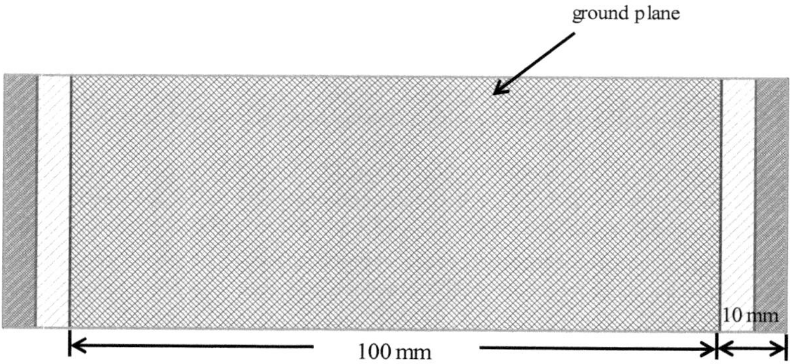


(b)

Figure 6.5: Top and bottom view of the two-port MIMO antenna with a double-stub matching circuit.

## 6.3 Different Two-Port MIMO Antenna Configurations

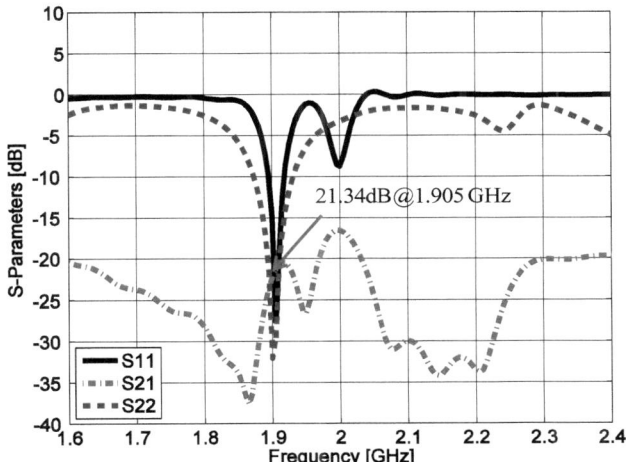

Figure 6.6: S-parameters of the two-port MIMO antenna with a double-stub matching circuit.

In the slotted ground configuration, creating slots in the ground plane easily match to the 50 Ω impedance through modifying the current distributions of the chassis modes. Slots lengths, widths, and positions are optimized so that the odd and even modes resonances are matched. For comparison, the current distribution of the odd chassis mode with and without a slot created at the middle of the chassis long end is shown in Fig.6.7. Creating a slot changes the current path for the odd mode, the mode resonance becomes a function of the slot length. For the even chassis mode, its resonance is not a function of the slot length at the middle of the chassis long end, while the odd and even modes resonances are a function of the slot length at the middle of the chassis short end. In the two-port MIMO antenna with a double-stub matching circuit, the effect of slots is replaced by a multiple-resonances external matching circuit.

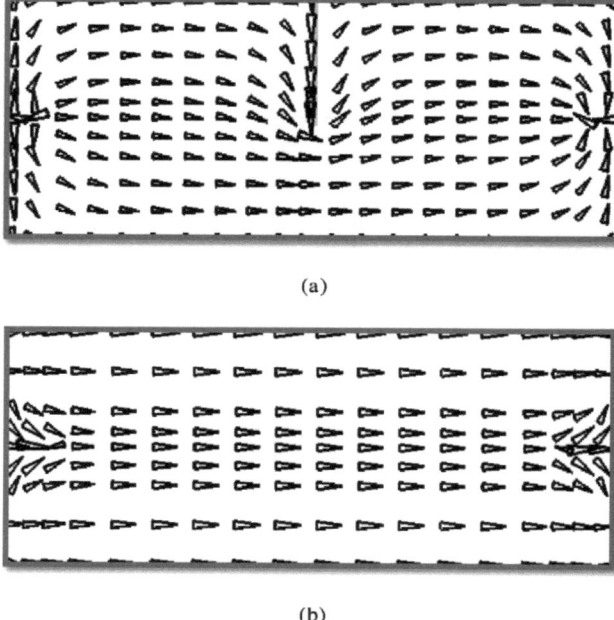

Figure 6.7: Current distribution of the odd chassis mode. (a) With and (b) without a slot at the middle of the chassis long end. (Arrows indicate the current direction).

### 6.3.3 Two-Port MIMO Antenna with an Offset Feed Point

In the offset fed two-port MIMO antenna, the feed point location for the CCEs is at 12.63 mm offset from the long axis of the chassis as shown in Fig.6.8a. The CCEs are off-ground elements with a ground clearance of 7 mm. The dimensions of the whole structure are 114×40×5.5 mm$^3$. A single-stub matching circuit is designed to match the ports to the 50 Ω impedance. The top and bottom view of the two-port MIMO antenna with an offset feed and a single-stub matching circuit is shown in Fig.6.8. The single-stub position and length for each port, the ground clearance of the CCEs, the CCEs height, and the offset of the feed point location from the chassis long axis are all optimized for good matching and port isolation at 1.83 GHz. An isolation of more than

## 6.3 Different Two-Port MIMO Antenna Configurations

17.9 dB is obtained over a frequency band from 1.80 to 1.858 GHz (6 dB bandwidth), Fig.6.9.

The CCEs with an offset feed point excite the half-wavelength mode of the chassis minor axis. The resonant frequency of the chassis minor axis half-wavelength mode was found at 2.7 GHz in [11]. In the offset feed point configuration, instead of a multiple-resonance matching circuit, an offset feed point (excites another resonance at 2.7 GHz) is used and then a single-resonance matching circuit is required.

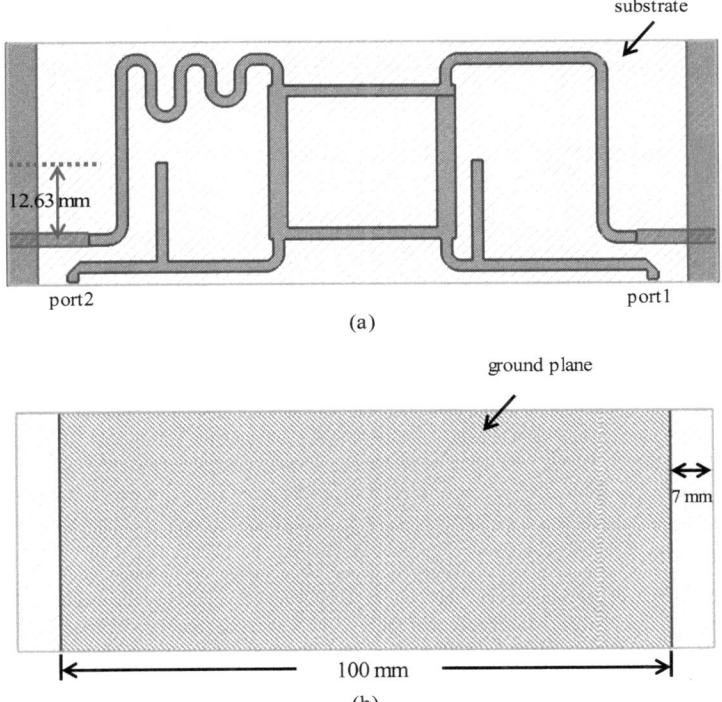

Figure 6.8: Top and bottom view of the two-port MIMO antenna with an offset feed and a single-stub matching circuit.

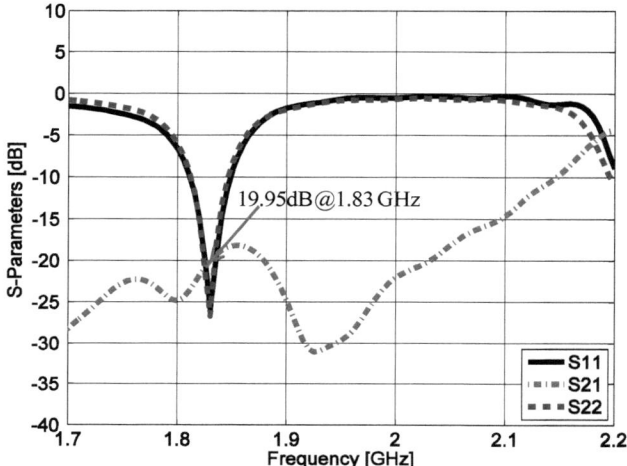

Figure 6.9: S-parameters of the two-port MIMO antenna with an offset feed and a single-stub matching circuit.

In the two-port MIMO antenna configurations discussed above, a narrow frequency bandwidth is obtained. This can be explained by the low input resistance of the coupling element and by the relative bandwidth calculated in Fig.1.11. The best bandwidth potential seems to be below 1 GHz, where the CCE is electrically small and the chassis dominates the radiation properties of the antenna. At around 2.2 GHz, the contribution of the chassis becomes less, while lower bandwidth potential is obtained between the two peaks where the design frequencies 1.83 and 1.90 GHz of the two-port MIMO antenna configurations discussed above are located.

For the relative bandwidth obtained in Fig.1.11, the CCE element is matched by an external single-resonant L-section matching circuit at each simulated frequency point between 0.5 GHz and 3 GHz. Using a complex (multiple-resonance) matching circuit enhances the bandwidth of the coupling-element based antenna structure. The relative bandwidth (10 dB bandwidth) as a function of the number of lumped elements in the matching circuit plotted in Fig.1.12 at 0.9 and 2.2 GHz shows that the relative bandwidth depends on the number of lumped elements in the external matching circuit. In addition, bandwidth enhancement could be obtained by optimizing the CCE

## 6.3 Different Two-Port MIMO Antenna Configurations

dimensions in comparison to the chassis dimensions for the purpose of high input resistance of the CCE.

Uncorrelated channels from the two-port MIMO antenna configurations mentioned above are obtained. The correlation coefficient can be calculated from the three dimensional complex far field radiation patterns or from the scattering parameters if an isotropic environment is assumed. In [84], the envelope correlation coefficient is derived and computed from the scattering parameters of the system as

$$\rho_e = \frac{\left|S_{11}^* S_{12} + S_{21}^* S_{22}\right|^2}{\left(1-|S_{11}|^2-|S_{21}|^2\right)\left(1-|S_{22}|^2-|S_{12}|^2\right)} \quad (6.6)$$

The envelope correlation coefficient calculated from (6.6) for the two-port MIMO antenna with a slotted ground configuration is $1.067 \times 10^{-4}$. The envelope correlation coefficient for the two-port MIMO antenna with a double-stub matching circuit configuration is $9.157 \times 10^{-7}$, and $1.813 \times 10^{-5}$ for the offset feed point configuration.

The radiation patterns of the half- and full-wavelength chassis modes for the two-port MIMO antenna with a slotted ground, a double-stub matching circuit, and an offset feed point configuration are shown in Fig.6.10a, b and c respectively. With respect to the E-plane (xz-plane in Fig.6.1), creation of slots in the ground plane does not any more keep the symmetry for the odd and even chassis modes, while for the H-plane (xy-plane in Fig.6.1), the odd mode pattern becomes more directive.

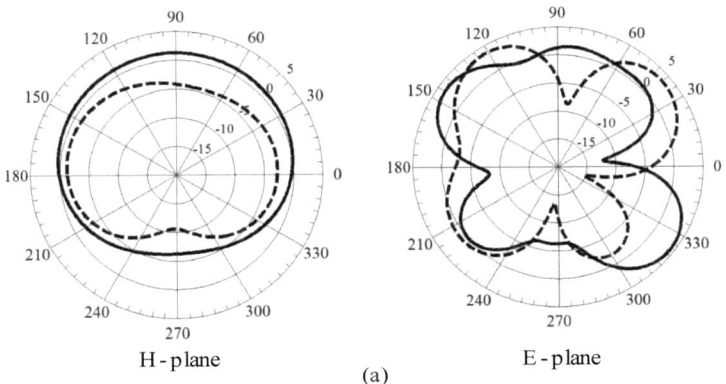

(a)

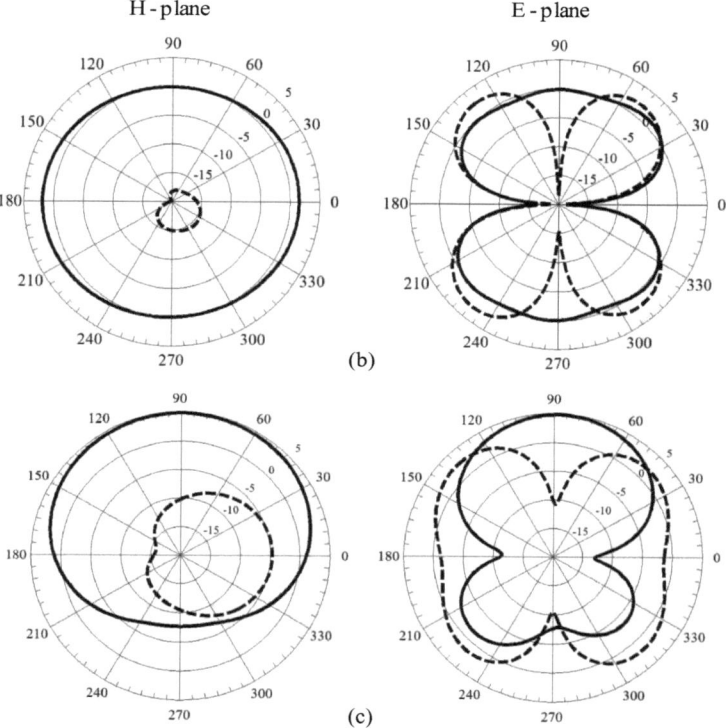

Figure 6.10: Radiation patterns of the odd (solid) and even (dash) chassis modes for the two-port MIMO antenna configurations. (a) With a slotted ground plane, (b) with a double-stub matching circuit, and (c) with an offset feed point.

For the offset feed point configuration, coupling to the half-wavelength mode of the chassis minor axis partially fills the nulls at $\theta = \pm 90°$ and fully fills the nulls at $\theta = 0°$ and $180°$ for the E-plane pattern of the even chassis mode. The odd chassis mode pattern in E-plane is a little bit distorted and becomes more directive in H-plane.

## 6.4 Eigenmodal Feed Based Decoupling Network of the MFSAA

For the MFSAA, four quarter-wavelength monopole elements are placed on a 100 mm × 100 mm ground plane. Due to the symmetry properties of the structure, the S-matrix of the MFSAA can be decomposed into

$$S = U\Gamma U^T$$
$$\Gamma = diag(\Gamma_1, \Gamma_2, \Gamma_3, \Gamma_4), \quad (6.7)$$

where S is the scattering matrix of the MFSAA, $\Gamma$ is the diagonal matrix

$$\begin{aligned}
\Gamma_1 &= S_{11} + S_{12} + S_{13} + S_{14} \\
\Gamma_2 &= S_{11} + S_{12} - S_{13} - S_{14} \\
\Gamma_3 &= S_{11} - S_{12} - S_{13} + S_{14} \\
\Gamma_4 &= S_{11} - S_{12} + S_{13} - S_{14},
\end{aligned} \quad (6.8)$$

and the unitary matrix $U$ consists of the orthonormal eigenvectors of S

$$U = \frac{1}{2}\begin{bmatrix} 1 & 1 & 1 & 1 \\ 1 & 1 & -1 & -1 \\ 1 & -1 & -1 & 1 \\ 1 & -1 & 1 & -1 \end{bmatrix}. \quad (6.9)$$

Exciting with the eigenvectors of the MFSAA, due to their orthogonality, isolated ports but mismatched are obtained. External matching networks are required to match each port to the 50 $\Omega$ impedance individually. The modal feed network that is required to generate the orthogonal eigenmodes of the MFSAA should have the scattering matrix $S^m$ described in a block matrix notation by

$$S^m = \begin{bmatrix} 0 & U^T \\ U & 0 \end{bmatrix}, \quad (6.10)$$

where $U$ is an $n \times n$ matrix ($n$ is the number of array elements, $n = 4$ for the MFSAA). The eigenmodal feed network of the MFSAA can be implemented by using four -3 dB 90° branchline couplers. In [83], the eigenmodal feed based decoupling network was realized on FR4 ($\varepsilon_r = 4.4$) substrate for a planar array of four monopole elements separated by $0.17\lambda$ at 2.6 GHz. The upper metallization of the substrate acts as the ground plane for the monopole elements, while the microstrip lines were etched on the bottom surface of the substrate. For pattern calculation an infinite ground was assumed (effect of chassis modes is excluded).

For the MFSAA, the monopole elements and the microstrip lines are modeled on the same surface (the upper surface) of $160\,\text{mm} \times 160\,\text{mm}$ RO4003 substrate of thickness 0.813 mm, dielectric constant $\varepsilon_r = 3.38$, and loss tangent 0.0027. The ground metallization surface is the bottom surface of the substrate. The MSL layout of the eigenmodal feed based decoupling matching network for the MFSAA is shown in Appendix F.

A single-stub external matching circuit for modes 1, 2, and 3 is realized, while a double-stub matching circuit is realized to match for mode 4. The scattering parameters of the MFSAA with an eigenmodal feed based decoupling network are shown in Fig.6.11.

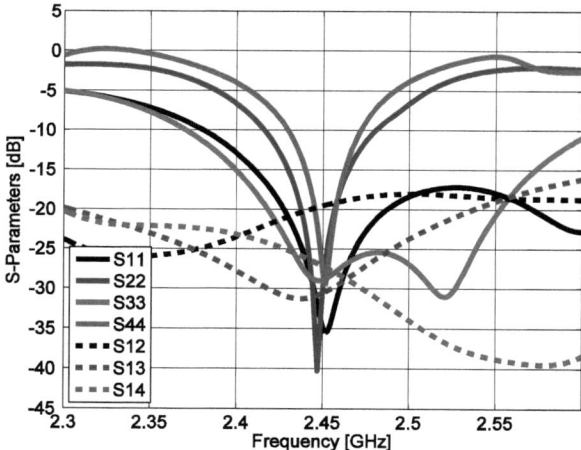

Figure 6.11: S-parameters of the MFSAA with an eigenmodal feed based decoupling and matching network.

## 6.4 Eigenmodal Feed Based Decoupling Network of the MFSAA

An isolation of more than 19.52 dB at 2.45 GHz is obtained. Wider frequency bandwidth (10 dB bandwidth) is obtained for modes 1 and 3 with excitation vectors [1 1 1 1] and [1 -1 -1 1] respectively compared to modes 2 and 4. The azimuth radiation patterns of the MFSAA eigenmodes 1, 2, 3 and 4 are shown in Fig.6.12.

If a finite ground is considered, the chassis characteristic modes fill the nulls of the MFSAA eigenmode azimuth patterns as can be seen in Fig.6.13. The layout dimensions ($1.3\lambda \times 1.3\lambda$) of the eigenmodal decoupling network are relatively large compared to the dimensions of the monopole array. For a smaller chassis compared to the monopole array size, the nulls of the array eigenmode azimuth patterns are expected to be completely filled.

In eigenmodal based decoupling networks, for a finite ground plane, nonresonant coupling element is required to excite the chassis modes and then port isolation is obtained based on the orthogonality of chassis modes radiation patterns. While for an infinite ground plane, port isolation is achieved from the orthogonality of array (self-resonant elements) eigenmode radiation patterns.

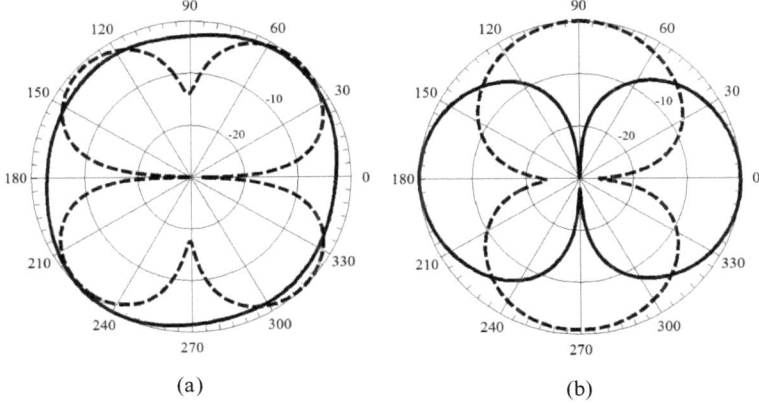

Figure 6.12: Azimuth radiation patterns of the MFSAA on an infinite ground plane. (a) For eigenmodes 1 (solid) and 4 (dash), and (b) for eigenmodes 2 (solid) and 3 (dash).

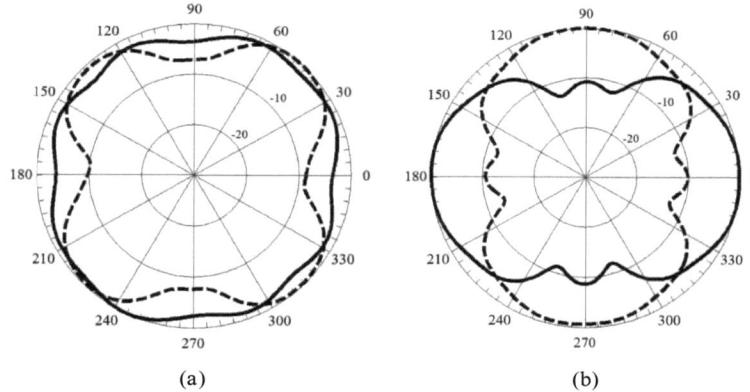

Figure 6.13: Azimuth radiation patterns of the MFSAA on a finite ground plane. (a) For eigenmodes 1 (solid) and 4 (dash), and (b) for eigenmodes 2 (solid) and 3 (dash).

# Chapter 7

# Prototypes: Fabrication and Measurements

## 7.1 Introduction

The present chapter describes design procedure, fabrication process, and measurements of the decoupling and matching network prototypes which were proposed in ch.5 and ch.6. Three different prototypes of decoupling and matching networks were designed and fabricated. The isolation between ports is measured. Comparison between measured and simulated results for each prototype and between different prototypes of networks is shown and discussed. Two prototypes of decoupling networks, parasitic decoupling elements and eigenmode based decoupling network, were fabricated to compensate for coupling degradation in the MFSAA. Another eigenmodal feed based decoupling prototype was fabricated for a two-port MIMO antenna. In this network, port isolation is obtained through coupling to the orthogonal radiation patterns of the chassis wavemodes.

## 7.2 Prototype I: Parasitic Elements Based Decoupling Technique of the MFSAA

The parasitic decoupling elements and matching circuit of the MFSAA are fabricated on a ground plane surface of $90\text{mm} \times 90\text{mm}$ ($0.735\lambda \times 0.735\lambda$), made from RO4003 substrate of thickness 0.813 mm, dielectric constant $\varepsilon_r = 3.38$, and loss tangent 0.0027. The array active elements are made from copper wires of $0.025\lambda$ in diameter, while the array parasitic elements are made from copper wires of $0.016\lambda$ in diameter. The decoupled ports are matched to the $50\,\Omega$ impedance with an open-circuit stub of length $0.23\lambda$ and width $0.01\lambda$. The stub is placed at distance $0.1\lambda$ from the feed point. For better matching another open-circuit stub of length $0.06\lambda$ and width $0.02\lambda$ is added. The fabricated whole structure of the MFSAA with parasitic decoupling elements and matching circuit is shown in Fig.7.1. The whole structure is placed on a large ground plane to provide an approximation of an infinite ground. Measured scattering parameters in Fig.7.2 show an isolation of more than 30 dB over a frequency band from 2.406 to 2.506 GHz (10 dB bandwidth). Good agreement between measured and simulated scattering parameters of the MFSAA is obtained. The scattering parameters $S_{12}$ and $S_{13}$ represent the coupling between adjacent and nonadjacent array active elements, respectively. For comparison, measured scattering parameters of the MFSAA without a decoupling and matching network are shown in Fig.7.3.

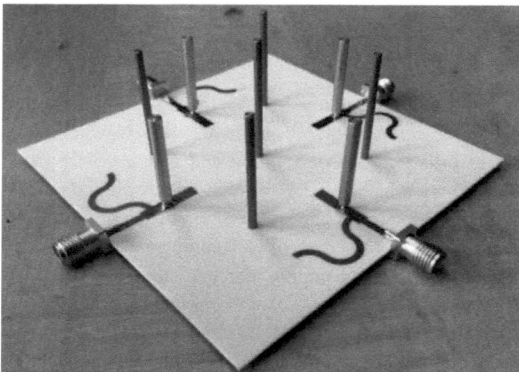

Figure 7.1: MFSAA with parasitic decoupling elements and matching circuit.

## 7.2 Prototype I: Parasitic Elements Based Decoupling Technique of the MFSAA 115

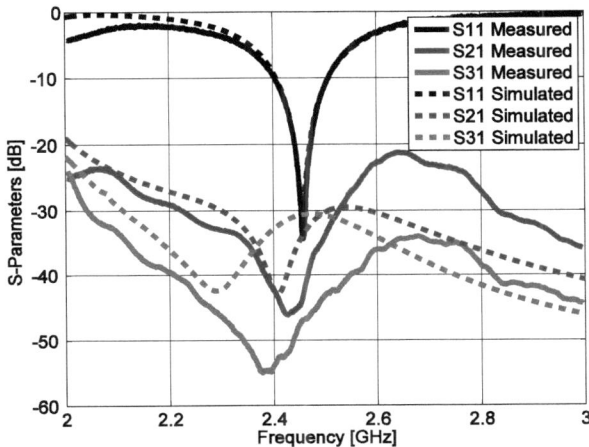

Figure 7.2: Measured and simulated S-parameters of the MFSAA with parasitic decoupling elements and matching circuit.

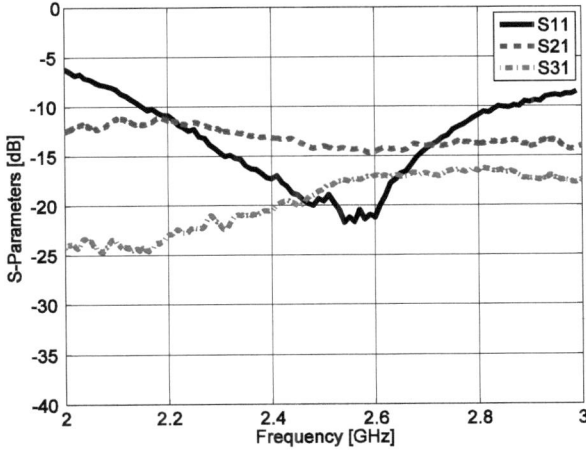

Figure 7.3: Measured S-parameters of the MFSAA without a decoupling network.

## 7.3 Prototype II: Eigenmode Based Decoupling Network of the MFSAA

The eigenmode based decoupling network of the MFSAA is realized in microstrip line and fabricated on a ground plane surface of $78\text{mm} \times 78\text{mm}$ ($0.637\lambda \times 0.637\lambda$) made from RO4003 substrate of thickness 0.508 mm, dielectric constant $\varepsilon_r = 3.38$, and loss tangent 0.0027. The array elements are made from copper wires of $0.025\lambda$ in diameter. The decoupled ports are matched to the 50 Ω impedance with an open-circuit double-stub of lengths $0.02\lambda$ and $0.06\lambda$, respectively from the feed point. The distance between the two stubs is $0.016\lambda$. The fabricated whole structure of the MFSAA with the decoupling and matching network is shown in Fig.7.4. The whole structure is placed on a large ground plane to provide an approximation of an infinite ground. Measured scattering parameters in Fig.7.5 show an isolation of more than 20 dB over a narrower frequency band compared to the parasitic decoupling elements. In addition, the scattering parameters $S_{12}$ and $S_{13}$ are a little bit shifted from $S_{11}$ compared to the simulated scattering parameters.

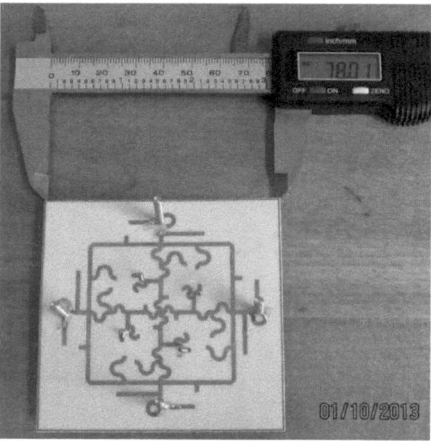

Figure 7.4: MFSAA with an eigenmode based decoupling network realized in microstrip line.

## 7.3 Prototype II: Eigenmode Based Decoupling Network of the MFSAA

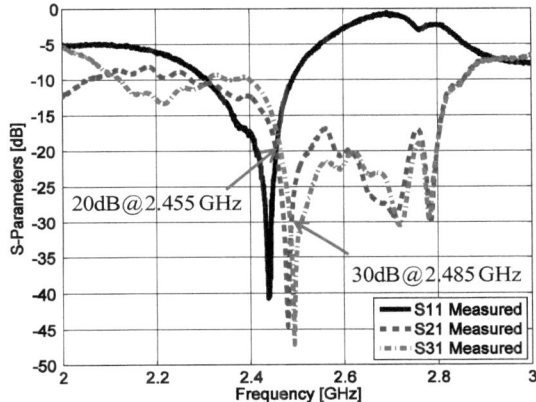

Figure 7.5: Measured S-parameters of the MFSAA with an Eigenmode based decoupling matching network realized in microstrip line.

To measure the radiation pattern of the MFSAA in Fig.7.4, a feed network is needed to generate the excitation vector of (4.4). Three branchline couplers are required to get uniform amplitudes and $90°$ phase shifts between output ports [61]. The feed network is fabricated on a ground plane surface of ($0.637\lambda \times 0.561\lambda$) made from RO4003 substrate of thickness 0.813 mm as shown in Fig.7.6.

Figure 7.6: Realization of the feed network for the MFSAA in Fig.7.4.

The measured S-parameters of the feed network are shown in Fig.7.7. Good return loss (-23.86 dB) and isolation (-26.89 dB) are obtained at 2.45 GHz. The transmission S-parameter phases are shown in Fig.7.8.

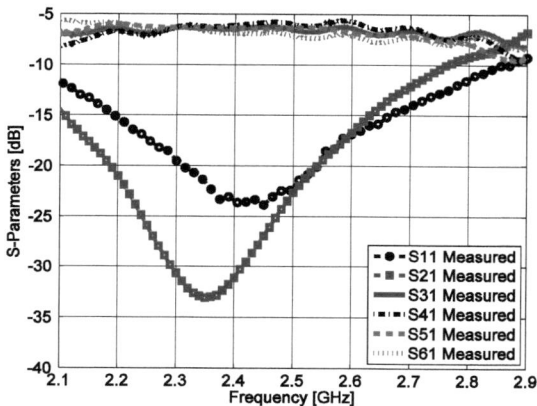

Figure 7.7: Measured S-parameter magnitudes of the feed network in Fig.7.6.

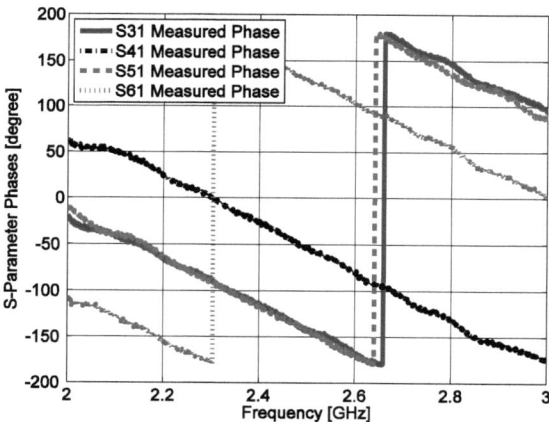

Figure 7.8: Measured transmission S-parameter phases of the feed network in Fig.7.6.

## 7.3 Prototype II: Eigenmode Based Decoupling Network of the MFSAA

A phase shift of 87.76° between port 4 and ports 3 and 5 is obtained, while a phase shift of 183.7° between port 4 and port 6 is obtained. The feed network is connected to the MFSAA with an eigenmode feed based decoupling network and the radiation pattern is measured in our anechoic chamber as shown in Fig.7.9. Simulated and measured azimuth patterns of the MFSAA are in a good agreement as shown in Fig.7.10. Ripples in the measured pattern are assumed to be due to superposition of the MFSAA radiation by spurious radiation from the cables and from the open feed network.

Figure 7.9: Pattern measurement of the MFSAA with the feed network of Fig.7.6.

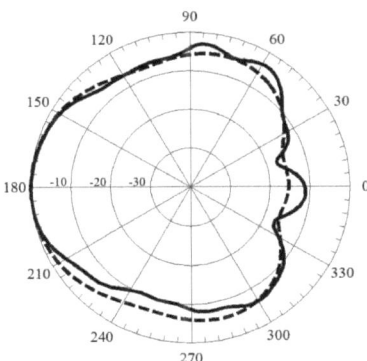

Figure 7.10: Measured (solid) and simulated (dash) azimuth pattern of the MFSAA.

For the purpose of beamforming in $\phi = 45°$ and nulls in $\phi = 135°, 225°,$ and $315°$ directions, the excitation vector of the MFSAA is optimized in section 5.3.2. For the optimized excitation vector in (5.38), one 11.06 dB attenuator and two 9.63 dB attenuators are realized by resistor Pi-section circuits. The phase shift between nonadjacent antenna elements in the MFSAA is adjusted to be $155°$ by insertion of a delay line. The whole structure is fabricated on a ground plane of ($0.327\lambda \times 0.639\lambda$) made from RO4003 substrate of thickness 0.508 mm as shown in Fig.7.11. The measured transmission S-parameters for the 9.63 dB attenuator, 11.06 dB attenuator and phase shifter are shown in Fig.7.12.

The measured and simulated azimuth patterns of the MFSAA with the eigenmode based decoupling network and excitation vector of (5.38) are shown in Fig.7.13. Strong spurious radiation superimposes and distorts the original MFSAA pattern, partially filling the nulls at $\phi = 90°$ and $\phi = 270°$ and producing ripples in the backward direction. In additions the null at $\phi = 270°$ is a little bit shifted in comparison to the original pattern. The Azimuth pattern of the MFSAA is measured at 2.48 GHz. From Fig.7.5, an isolation of 30 dB is obtained at 2.485 GHz.

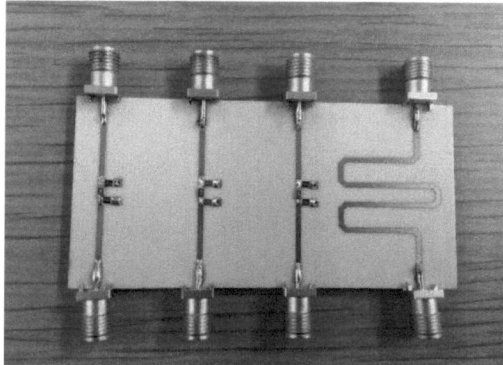

Figure 7.11: Realization of the attenuators and phase shifter for the excitation vector in (5.38).

## 7.3 Prototype II: Eigenmode Based Decoupling Network of the MFSAA 121

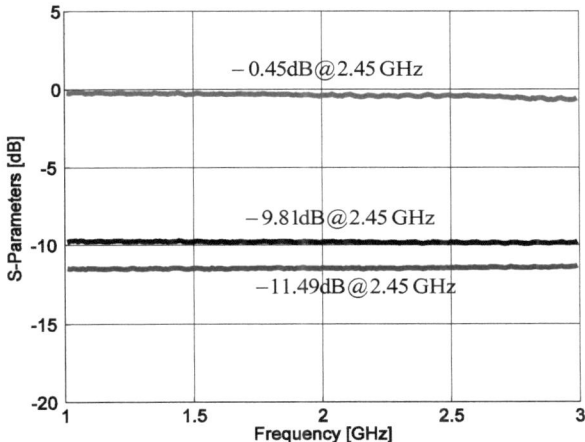

Figure 7.12: Measured transmission S-parameters of the structure in Fig.7.11.

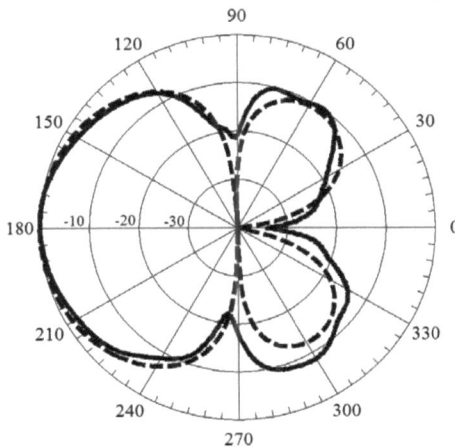

Figure 7.13: Measured (solid) and simulated (dash) azimuth pattern of the MFSAA with the excitation vector in (5.38).

## 7.4 Prototype III: Eigenmodal Feed Based Decoupling Network of a Two-Port MIMO Antenna

The first two configurations with and without a slotted ground plane of a two-port MIMO antenna modeled in ch.6 are fabricated on RO4003 substrate of thickness 0.813mm, dielectric constant $\varepsilon_r = 3.38$, and loss tangent 0.0027. The CCE are made from copper plate of thickness 100 µm. The decoupled ports of the two-port MIMO antenna with a slotted ground configuration are matched to the 50 Ω impedance with a single-stub matching circuit, while for the two-port MIMO antenna without a slotted ground, a double-stub matching circuit is used. The fabricated whole structure and measured scattering parameters for the two-port MIMO antenna with a slotted ground and a single-stub matching circuit are shown in Fig.7.14 and Fig.7.15, respectively. While Fig.7.16 and Fig.7.17 show the fabricated whole structure and measured scattering parameters of the two-port MIMO antenna with a double-stub matching circuit. An isolation of around 20dB is obtained for the two-port MIMO antenna with a double-stub matching circuit. The measured $S_{22}$ and $S_{21}$ are in a good agreement with the simulated ones.

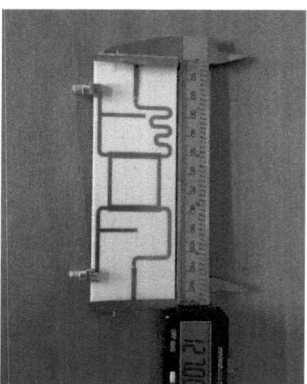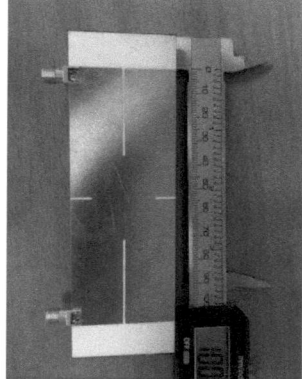

Figure 7.14: Top and bottom view of the two-port MIMO antenna with a slotted ground and a single-stub matching circuit.

## 7.4 Prototype III: Eigenmodal Feed Based DN of a Two-Port MIMO Antenna

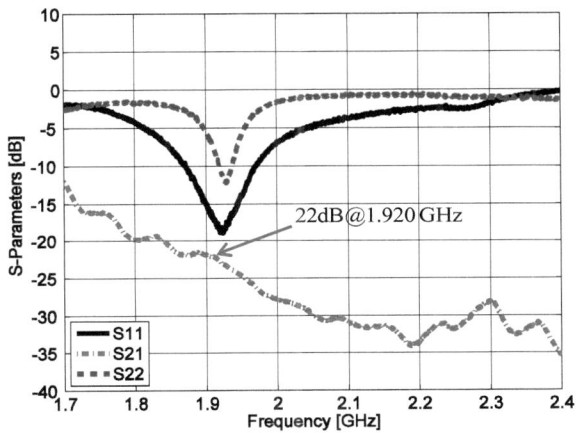

Figure 7.15: Measured S-parameters of the two-port MIMO antenna with a slotted ground and a single-stub matching circuit.

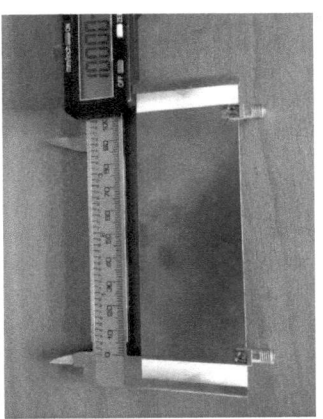 

Figure 7.16: Top and bottom view of the two-port MIMO antenna with a double-stub matching circuit.

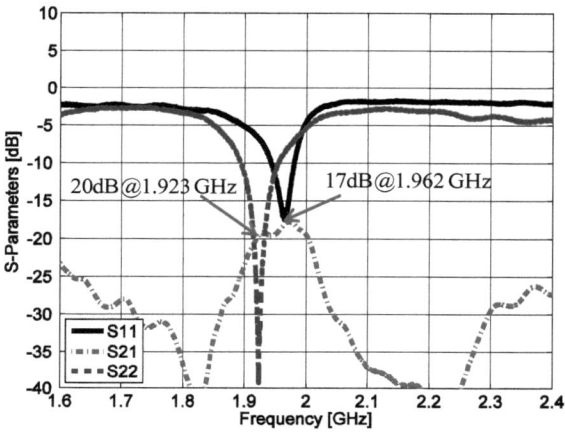

Figure 7.17: Measured S-parameters of the two-port MIMO antenna with a double-stub matching circuit.

The impedance match frequency for $S_{11}$ is a little bit shifted from 1.923 GHz due to loading at the chassis ends with the coaxial cables for feeding the structure, where the odd chassis mode is strongly coupled. From the measured scattering parameters in Fig.7.17, the frequency band (6 dB bandwidth) is between 1.914 and 1.992 GHz for the $1^{st}$ chassis mode (odd mode) and between 1.863 and 1.991 GHz for the $2^{nd}$ chassis mode (even mode).

For the slotted ground plane configuration, an isolation of more than 20 dB is obtained over a wider frequency band compared to the two-port MIMO antenna with a double-stub matching circuit. The shift in the resonant frequency for $S_{11}$ is compensated by modifying the external matching circuit for port1 to be a double-stub matching circuit instead of a single-stub. With a double-stub matching circuit for port1 in addition to slots in the ground plane, a wider frequency band from 1.832 to 2.018 GHz (6 dB bandwidth) is obtained compared to a narrower frequency band from 1.904 to 1.950 GHz for port2. The matched frequency band at -6 dB level from simulated and measured S-parameters for the two configurations of the two-port MIMO antenna are summarized in Tab. 7.1.

## 7.4 Prototype III: Eigenmodal Feed Based DN of a Two-Port MIMO Antenna

TABLE 7.1: MATCHED FREQUENCY BAND IN GHZ FOR THE TWO CONFIGURATIONS OF THE TWO-PORT MIMO ANTENNA.

| Matched frequency band in GHz at -6 dB level | Simulated S-Parameters | | Measured S-Parameters | |
|---|---|---|---|---|
| | port1 | port2 | port1 | port2 |
| Two-port MIMO antenna with a slotted ground | 1.899- 1.948 | 1.898- 1.948 | 1.832- 2.018 | 1.904- 1.950 |
| Two-port MIMO antenna with a double-stub matching circuit | 1.887-1.925 | 1.852-1.954 | 1.914- 1.992 | 1.863- 1.991 |

# Chapter 8

## Conclusion and Future Work

Multiport antennas usually have the design goal of isolated ports and uncorrelated radiation patterns. For antennas on small platforms (e.g. mobile terminals), spacing of less than half-wavelength may result. Small spacing between antenna elements leads to a strong mutual coupling and correlated radiation patterns. In addition, for antennas on small platforms, the characteristic chassis modes effect has to be considered. The chassis is the main radiator especially at low frequencies, while the antenna element works as an exciter to excite the chassis wavemodes. Obviously, a good knowledge of the characteristic modes of the chassis is so helpful to analyze the radiation properties of the chassis of a mobile terminal.

In ch.3, for deep understanding of how different parts of the antenna-chassis combination contribute to the input impedance and total radiation, a lumped elements equivalent circuit model based on parallel and series coupled resonators to model the chassis wavemodes and the antenna elements wavemodes is presented. By the help of the equivalent circuit model, the effect of chassis wavemodes on the input impedance and their contribution to the total radiation can be studied.

Mutual coupling in the MFSAA modifies the monopole input impedances, and accordingly the terminal voltages are modified resulting in a modified array pattern. The distortion of the radiation pattern of the MFSAA can be compensated by modifying the feeding network of the MFSAA. On the other hand, mutual coupling between array elements reduces the signal to noise ratio of the system (SNR). Modification of excitation vector in beamforming antenna arrays cannot compensate for SNR degradation. In such case, implementation of a decoupling network between the input ports and antenna ports is essential for high SNR. Different decoupling techniques for the MFSAA are presented in ch.5. Based on the eigenmode theory, for the MFSAA, two stages of a lossless decoupling network are required. In the $1^{st}$ stage, modes (1, 4) admittances are matched, while in the $2^{nd}$ stage modes (1, 4) and modes (2, 3) are matched. All ports are now matched over a narrow frequency band. An isolation of more than 20 dB is obtained over a frequency band from 2.427 GHz to 2.453 GHz.

To decouple over a wider frequency band, an alternative approach is presented. By inserting parasitic elements between adjacent and nonadjacent array active elements, good isolation of more than 30 dB between antenna ports of the MFSAA is obtained over a wider frequency band from 2.406 GHz to 2.506 GHz (10 dB bandwidth). For both techniques of decoupling, considering the effect of chassis modes, the H-plane pattern of the half-wavelength chassis mode fills the null at $\theta = 0°$. The E-plane patterns of the half- and full-wavelength chassis modes are responsible for the distortion in the azimuth pattern of the MFSAA.

The theory of characteristic modes is utilized in ch.6 to get port isolation by making use of the orthogonality of characteristic wavemodes of the chassis. A two-port MIMO antenna is proposed with two CCEs to excite the first two dominants characteristic modes of a 100 mm × 40 mm chassis. The CCEs shape and location have to be properly chosen for strong coupling to the $1^{st}$ and $2^{nd}$ chassis modes. Different configurations of a two-port MIMO antenna with and without modified ground plane and offset feed point are presented and port isolation over a narrow frequency band is obtained. For the two-port MIMO antenna with a double-stub matching circuit, the odd chassis mode is matched over a frequency range between 1.887 GHz and 1.925 GHz (at the -6 dB level) of more than 20 dB isolation, while the even chassis mode is matched over a wider frequency band from 1.852 GHz to 1.954 GHz. For the two-port MIMO antenna with a slotted ground plane and a single-stub matching circuit, the odd and even chassis modes are matched at the same frequency band from 1.898 GHz to 1.948 GHz with isolation of more than 25 dB, while the odd and even chassis modes for the two-port MIMO antenna with an offset feed point are matched at the same frequency band from 1.8 GHz to 1.858 GHz with isolation of more than 18 dB.

For the radiation patterns of the half- and full-wavelength chassis modes, creation of slots in the ground plane gives up the symmetry for the odd and even chassis modes in E-plane, while for H-plane, the odd mode pattern becomes more directive. In the offset feed configuration, coupling to the half-wavelength mode of the chassis minor axis partially fills the nulls at $\theta = \pm90°$ and fully fills the nulls at $\theta = 0°$ and $180°$ for the even chassis mode pattern in E-plane. The odd chassis mode pattern in E-plane is a little bit distorted.

In ch.7, different decoupling network prototypes for the MFSAA and different two-port MIMO antenna configurations are fabricated and good agreement between measured and simulated results is obtained.

For future work, the two-port MIMO antenna with an eigenmodal feed based decoupling network can be extended to a four-port MIMO antenna for higher channel capacity. The diversity performance of two- or more port antennas with different decoupling techniques can be evaluated by calculating the envelope correlation coefficient taking into account the propagation environment. For isotropic, the envelope correlation coefficient can be calculated from the scattering parameters.

# REFERENCES

[1] J. H. Winters, J. Salz and R. D. Gitlin, "The impact of antenna diversity on the capacity of wireless communication systems", *IEEE Trans. Communications*, vol. 42, no. 2/3/4, pp. 1740-1751, Feb/Mar/Apr 1994.

[2] I. J. Gupta and A. A. Ksienski, "Effect of mutual coupling on the performance of adaptive arrays", *IEEE Trans. Antennas and Propagation*, vol. 31, no. 5, pp. 785-791, September 1983.

[3] R. J. Garbacz, and R. H. Turpin, "A generalized expansion for radiated and scattered fields," *IEEE Trans. Antennas Propagat.*, vol. AP-19, no. 3, pp. 348-358, May 1971.

[4] R. F. Harrington, and J. R. Mautz, "Theory of characteristic modes for conducting bodies," *IEEE Transactions on Antennas and Propagation*, vol. AP-19, no. 5, pp. 622-628, September 1971.

[5] R. F. Harrington, *Field Computation by Moment Methods*, New York, MacMillan, 1968.

[6] R. F. Harrington, and J. R. Mautz, "Computation of characteristic modes for conducting bodies," *IEEE Transactions on Antennas and Propagation*, vol. AP-19, no. 5, pp. 868-871, September 1971.

[7] E. H. Newman, "Small antenna location synthesis using characteristic modes," *IEEE Transactions on Antennas and Propagation*, vol. AP-21, no. 6, pp. 629-639, November 1973.

[8] A. D. Yaghjian, and S. R. Best, "Impedance, bandwidth, and Q of antennas," *IEEE Transactions on Antennas and Propagation*, vol. 53, no. 4, pp. 1298-1324, April 2005.

[9] W. Geyi, P. Jarmuszewski, and Y. Qi, "The Foster reactance theorem for antennas and radiation Q," *IEEE Transactions on Antennas and Propagation*, vol. 48, no. 3, pp. 401-408, March 2000.

[10] C. Famdie, "Small antennas radiation performance optimization in mobile communications," Ph.D. Dissertation, Duisburg-Essen University, 2007.

[11] C. Famdie, W. Schroeder, and K. Solbach, "Numerical analysis of characteristic modes on the chassis of mobile phones," in *Proc. 1$^{st}$ EuCAP*, Nice, France, November 2006.

[12] G. J. Burke and A. J. Poggio, *Numerical Electromagnetics Code (NEC)*, 1981. www.nec2.org.

[13] E. Anderson et al., *LAPACK users' Guide*, 1999. www.netlib.org/lapack.

[14] P. Vainikainen, J. Ollikainen, O. Kivekäs, and I. Keander, "Resonator-based analysis of the combination of mobile handset antenna and chassis," *IEEE Transactions on Antennas and Propagation*, vol. 50, no. 10, pp. 1433-1444, October 2002.

[15] J. Villanen, J. Ollikainen, O. Kivekäs, and P. Vainikainen, "Coupling element based mobile terminal antenna structures," *IEEE Transactions on Antennas and Propagation*, vol. 54, no. 7, pp. 2142-2153, July 2006.

[16] J. Rahola, and J. Ollikainen, "Optimal antenna placement for mobile terminals using characteristic mode analysis," in *Proc. 1$^{st}$ EuCAP2006*, Nice, France, November 2006.

[17] C. T. Famdie, W. L. Schroeder, and K. Solbach, "Optimal antenna location on mobile phones chassis based on the numerical analysis of the characteristic modes," in *Proc. 37th EuMW2007*, Munich, October 2007.

[18] R. Valkonen, M. Kaltiokallio, and C. Icheln, "Capacitive coupling element antennas for multi-standard mobile handsets," *IEEE Transactions on Antennas and Propagation*, vol. 61, no. 5, pp. 2783-2791, May 2013.

[19] S. K. Chaudhury, H. J. Chaloupka, and A. Ziroff, "Multiport antenna systems for MIMO and Diversity," *EuCAP2010*, Barcelona, Spain, April 2010.

[20] R. Martens, E. Safin, and D. Manteuffel, "Selective excitation of characteristic modes on small terminals," in *Proc. EuCAP2011*, Rome, Italy, April 2011.

[21] R. Martens, E. Safin, and D. Manteuffel, "Inductive and capacitive excitation of the characteristic modes of small terminals," *Loughborough Antenna and Propagation Conference LAPC2011*, Loughborough, UK, November 2011.

# REFERENCES

[22] D. Manteuffel, and M. Arnold, "Considerations for reconfigurable multi-standard antennas for mobile terminals," in *Proc. Int. Workshop Antenna Technology (iWAT2008)*, Chiba, Japan, March 2008.

[23] D. M. Pozar, Microwave Engineering, 3$^{rd}$ Ed., John Wiley & Sons, Inc., USA, 2005.

[24] A. Andujar, J. Anguera, and C. Puente, "Ground plane boosters as a compact antenna technology for wireless handheld devices," *IEEE Trans. Antennas Propag.*, vol. 59, no. 5, pp. 1668-1677, May2011.

[25] R. Valkonen, J.Ilvonen, and P. Vainikainen, "Naturally non-selective handset antennas with good robustness against impedance mistuning," *EuCAP2012*, Prague, March 2012.

[26] J. Villanen, C. Icheln, and P. Vainikainen, "A coupling element based quad-band antenna structure for mobile terminals," *Microwave and Optical Technology Letters*, vol. 49, no. 6, pp. 1277-1282, June 2007.

[27] R. Valkonen, M. Kaltiokallio, and C. Icheln, "Capacitive coupling element antennas for multi-standard mobile handsets," *IEEE Transactions on Antennas and Propagation*, vol. 61, no. 5, May 2013.

[28] J. Villanen, M. Mikkola, C. Icheln, and P. Vainikainen, "Radiation characteristic of antenna structures in clamshell-type phones in wide frequency range," in *Proc. Vehicular Technology Conference VTC2007*, Dublin, Ireland, April 2007.

[29] R. Valkonen, J. Holopainen, C. Icheln, and P. Vainikainen, "Broadband tuning of mobile terminal antennas," in *Proc. EuCAP2007*, Edinburgh, UK, November 2007.

[30] R. Valkonen, C. Luxey, J. Holopainen, C. Icheln, and P. Vainikainen , "Frequency-reconfigurable mobile terminal antenna with MEMS switches," in *Proc. EuCAP2010*, Barcelona, Spain, April 2010.

[31] Z. H. Hu, J. Kelly, C. T. P. Song, P. S Hall, and P. Gardner , "Novel wide tunable dual-band reconfigurable chassis-antenna for future mobile terminals," in *Proc. EuCAP2010*, Barcelona, Spain, April 2010.

## 132 REFERENCES

[32] G. M. Rebeiz, "RF MEMS switches: status of the technology," *The 12$^{th}$ International Conference on Solid State Sensors, Actuator and Microsystems*, Boston, June 2003.

[33] R. Chan, R. Lesnick, D. Becher, and M. Feng, "Low-actuation voltage RF MEMS shunt switch with cold switching lifetime of seven billion cycles," *Journal of Microelectromechanical Systems*, vol. 12, no. 5, October 2003.

[34] A. Sutono, D. Heo, Y.-J. E. Chen, and J. Laskar, "High-Q LTCC-based passive library for wireless system-on-package (SOP) module development," *IEEE Trans. Microw. Theory Tech.*, vol. 49, no. 10, pp. 1715-1724, October 2001.

[35] O.-S. Lin, C.-C. Liu, K.-M. Li, and C. H. Chen, "Design of an LTCC tri-band transceiver module for GPRS mobile applications," *IEEE Trans. Microwave Theory Tech.*, vol. 52, no. 12, pp. 2718-2724, December 2004.

[36] U. Bulus, C. Famdie, and K. Solbach, "Equivalent-circuit modelling of the chassis radiator," *GeMIC2009*, Munich, Germany, March 2009.

[37] U. Bulus, and K. Solbach, "Modelling of the monopole interaction with a small chassis," in Proc. *EuCAP2009*, Berlin, Germany, March 2009.

[38] P. Yazdanpakhsh, and K. Solbach, "A circuit model of monopole four square array antenna on a finite ground plane including mutual coupling effects," *EuCAP2010*, Barcelona, Spain, April 2010.

[39] J. Holopainen, R. Valkonen, O. Kivekäs, J.Ilvonen, and P. Vainikainen, "Broadband equivalent circuit model for capacitive coupling element based mobile terminal antenna," *IEEE Antennas and wireless Propagation Letters*, vol. 9, pp.716-719 ,2010.

[40] S. Salama, and K. Solbach, "Equivalent Circuit Modeling of Monopoles on a Small Platform," *iWAT 2013*, Karlsruhe, Germany, March 2013.

[41] Z. H. Hu, J. Kelly, C. T. P. Song, P. S. Hall, and P. Gardner, "Equivalent circuit modeling of chassis antenna with two coupling elements," *Antennas and Propagation Society International Symposium (APSURSI)*, Toronto, July 2010.

# REFERENCES

[42] S. Salama, and K. Solbach, "Study of mutual coupling and chassis modes coupling through the equivalent circuit modeling of two monopoles on a small platform," *LAPC 2014*, Loughborough, UK, November 2014.

[43] K. Solbach, and C. T. Famdie, "Mutual coupling and chassis-mode coupling in small phased array on small ground plane," *EuCAP 2007*, Edinburgh, November 2007.

[44] H. Li, Y. Tan, B. Lau, Z. Ying, and S. He, "Characteristic mode based tradeoff analysis of antenna-chassis interaction for multiple antenna terminals," *IEEE Transactions on Antennas and Propagation*, vol. 60, no.2, February 2012.

[45] I. J. Gupta and A. A. Ksienski, "effect of mutual coupling on the performance of adaptive arrays," *IEEE Trans. Antennas Propagat.*, vol. AP-31, pp. 785-791, September 1983.

[46] B. Friedlander and A. J. Wiss, "Direction finding in the presence of mutual coupling," *IEEE Trans. Antennas Propagat.*, vol. 39, pp. 273-284, March 1991.

[47] K. M. Pasala and E. M. Friel, "Mutual coupling effects and their reduction in wideband direction of arrival estimation," *IEEE Trans. Aerosp. Electron. Syst.*, vol. 30, pp. 1116-1122, April 1994.

[48] D. F. Kelly and W. L. Stutzman, "Array antenna pattern modeling methods that include mutual coupling effects," *IEEE Transactions on Antennas and Propagation*, vol. 41, no. 12, December 1993.

[49] C. A. Balanis, Antenna Theory, 3$^{rd}$ Ed., John Wiley & Sons, Inc., Hoboken, New Jersey, 2005.

[50] J. D. Kraus and R. J. Marhefka, Antennas for all Applications, 3$^{rd}$ Ed., McGraw-Hill, New York, 2002.

[51] N. Guan, H. Furuya, D. Delaune, and k. Ito, "Radiation efficiency of monopole antenna made of a transparent conductive film," in *Proc. 2007 IEEE AP-S Int. Symp*, Hawaii, USA, June 2007.

[52] H. T. Hui, K. Y. Chan, and E. K. Yung, "Compensating for the mutual coupling effect in a normal-mode helical antenna array for adaptive nulling", *IEEE Transactions on Vehicular Technology*, vol. 52, no. 4, July 2003.

[53] H. T. Hui, "A practical approach to compensate for the mutual coupling effect in an adaptive dipole array", *IEEE Transactions on Antennas and Propagation*, vol. 52, no. 5, May 2004.

[54] H. T. Hui, "A new definition of mutual impedance for application in dipole receiving antenna arrays," *IEEE Antennas and Wireless Propagation Letters*, vol. 3, 2004.

[55] F. Yang, and Y. Rahmat-Samii, "Microstrip antennas integrated with electromagnetic band-gap (EBG) structures: a low mutual coupling design for array applications", *IEEE Transactions on Antennas and Propagation*, vol. 51, no. 10, October 2003.

[56] L. Li, B. Li, H. X. Liu, and C. H. Liang, "Locally resonant cavity cell model for electromagnetic band gap structures", *IEEE Trans. Antennas Propag.*, vol. 54, no. 1, pp. 90-100, January 2006.

[57] M. Karaboikis, C. Soras, G. Tsachtsiris and V. Makios, "Compact dual-printed inverted-F antenna diversity systems for portable wireless devices", *IEEE Antennas and Wireless Propagation letters*, vol. 3, 2004.

[58] M. Salehi, A. Motevasselian, A Tavakoli, and T. Heidari, "Mutual coupling reduction of microstrip antennas using defected ground structure", *10th IEEE Singapore International Conference on Communication systems*, October 2006.

[59] Q. Luo, H. M. Salgado, and J. R. Pereira, "Compact printed monopole antenna array for dual-band WLAN application", *EUROCON 2011 IEEE*, Lisbon, April 2011.

[60] I. Ahmad, W. F. Perger, and S. R. Zekavat, "Effects of ground constituent parameters on array mutual coupling for DOA estimation", *International Journal of Antennas and Propagation*, ID 425913, August 2011.

# REFERENCES

[61] P. Yazdanbakhsh, "Contributions of the performance optimization of the monopole four square array antenna," Ph.D. Dissertation, Duisburg-Essen University, 2011.

[62] H. J. Chaloupka, X. Wang, and J. Coetzee, "A Superdirective 3-element array for adaptive beamforming", *Microwave and Optical Technology Letters*, vol. 36, no. 6, March 2003.

[63] P. T. Chua, and J. C. Coetzee, "Microstrip decoupling networks for low-order multi-port arrays with reduced element spacing", *Microwave and Optical Technology Letters*, pp. 592-597, 2005.

[64] J. C. Coetzee, and Y. Yu, "Closed-form design equations for decoupling networks of small arrays", *Electronics Letters*, vol. 44, no. 25, December 2008.

[65] J. C. Coetzee, and Y. Yu, "Design of decoupling networks for circulant symmetric antenna arrays", *IEEE Antennas and Wireless Propagation Letters*, vol. 8, May 2009.

[66] S. Salama, and K. Solbach, "Design of decoupling network for monopole four square array antenna for multi-beam applications," *LAPC 2013*, Loughborough, UK, November 2013.

[67] A. Diallo, C. Luxey, P. L. Thuc, R. Straraj, and G. Kossiavas, "Enhancement of the isolation between two closely spaced mobile internal antennas by a neutralization effect", presented at the $22^{nd}$ *Int. Rev. Progr. Appl. Comput. Electromagn. (ACES 2006)*, Miami, FL, March 2006.

[68] V. Ssorin, A. Artemenko, A. Maltsev, A. Sevastyanov, and R. Maslennikov, "Compact MIMO microstrip antennas for USB dongle operating in 2.5-2.7 GHz frequency band", *International Journal of Antennas and Propagation*, ID 793098, August 2012.

[69] B. K. Lau, and J. B. Andersen, "Simple and efficient decoupling of compact arrays with parasitic scatterers", *IEEE Transactions on Antennas and Propagation*, vol. 60, no. 2, February 2012.

[70] A. C. K. Mak, C. R. Rowell, and R. D. Murch, "Isolation enhancement between two closely packed antennas", *IEEE Transactions on Antennas and Propagation*, vol. 56, no. 11, November 2008.

[71] S. Salama, and K. Solbach, "Parasitic Elements Based Decoupling Technique for Monopole Four Square Array Antenna", *EuMW2014*, Rome, Italy, October 2014.

[72] C. Calos, H. Okabe, T. Iwai, and T. Itoh, "A simple an accurate model for microstrip structures with slotted ground plane", *IEEE Microwave Wireless Comp. Lett.*, vol. 14, no. 4, pp. 133-135, April 2004.

[73] E. E. Altshuler, T. H. O'Donnell, A. D. Yaghjian, and S. R. Best, "A monopole superdirective array", *IEEE Transactions on Antennas and Propagation*, vol. 53, no. 8, August 2005.

[74] S. R. Best, E. E. Altshuler, A. D. Yaghjian, J. M. McGinthy, and T. H. O'Donnell, "An impedance-matched 2-element superdirective array", *IEEE Antennas and Wireless Propagation Letters*, vol. 7, 2008.

[75] T. H. O'Donnell, and A. D. Yaghjian, "Electrically small superdirective arrays using parasitic elements", *IEEE Antennas and Propagation Society International Symposium 2006*, Albuquerque, NM, July 2006.

[76] H. Carrasco, H. D. Hristov, R. Feick, and D. Cofre, "Mutual coupling between planar inverted-F antennas", *Microwave Opt. Technol. Lett.*, vol. 42, no. 3, pp. 224-227, August 2004.

[77] Y. A. Dama, and et al, "An envelope correlation formula for (N, N) MIMO antenna arrays using input scattering parameters, and including power losses," *International Journal of Antennas and Propagations*, ID 421691, vol. 2011.

[78] J. Bach. Andersen, and H. H. Rasmussen, "Decoupling and descattering networks for antennas", *IEEE Transactions on Antennas and Propagation*, November 1976.

[79] H. J. Chaloupka, and X. Wang, "Novel approach diversity and MIMO antennas at small mobile platforms", in *Proc. IEEE Int. Symp. On Personal, Indoor and Mobile Radio communications*, vol. 1, pp. 637-642, September 2004.

# REFERENCES

[80] J. C. Coetzee, "Dual-frequency decoupling networks for compact antenna arrays", *International Journal of Microwave Science and Technology*, ID 249647, 2011.

[81] S. Salama, and K. Solbach, "Eigenmodal Feed Based Decoupling Network for Two Ports MIMO and Diversity," *LAPC 2014*, Loughborough, UK, November 2014.

[82] S. K. Chaudhury, W. L. Schroeder, and H. J. Chaloupka, "MIMO antenna system based on orthogonality of the characteristic modes of a mobile device," in Proc. *2nd International ITG Conference on Antennas INICA2007*, Munich, Germany, March 2007.

[83] J. C. Coetzee, and Y. Yu, "Port decoupling for small arrays by means of an eigenmode feed network", *IEEE Transactions on Antennas and Propagation*, vol. 56, no. 6, June 2008.

[84] S. Blanch, J. Romeu, and I. Corbella, "Exact representation of antenna system diversity performance from input parameter description", *Electronics Letters*, vol. 39, no. 9, May 2003.

# APPENDICES

## APPENDIX A

**Equivalent Circuit Model in ADC of a Single Monopole on a 100 mm x 40 mm Chassis**

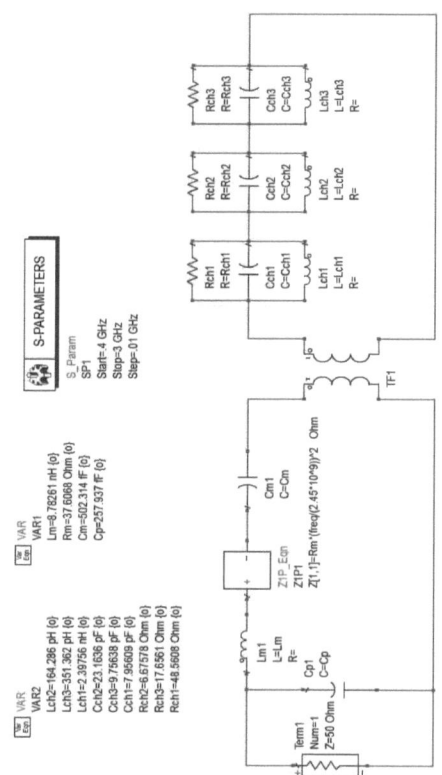

# APPENDIX B

Equivalent Circuit Model in ADC of Two Monopoles at the Middle of a 100 mm x 40 mm Chassis Short Ends

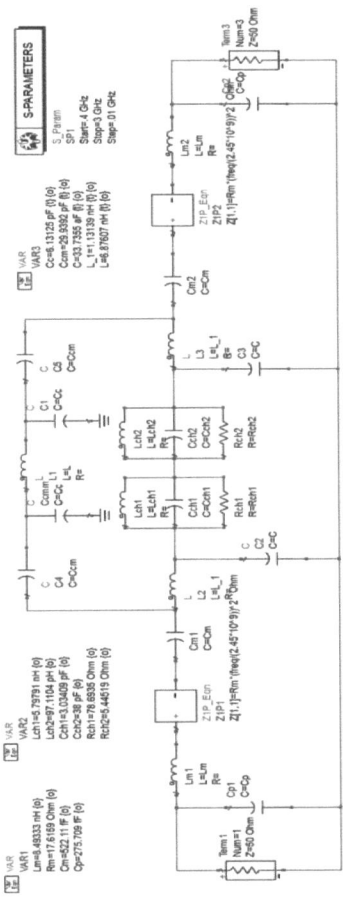

## APPENDIX C

**Equivalent Circuit Model in ADC of Two Monopoles at the Middle of a 100 mm x 40 mm Chassis Long Ends**

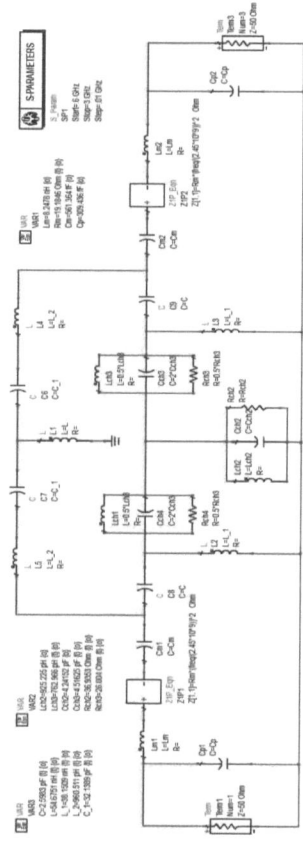

# APPENDIX D

Circuit Model in ADS for the Eigenmode Based Decoupling Network of the MFSAA

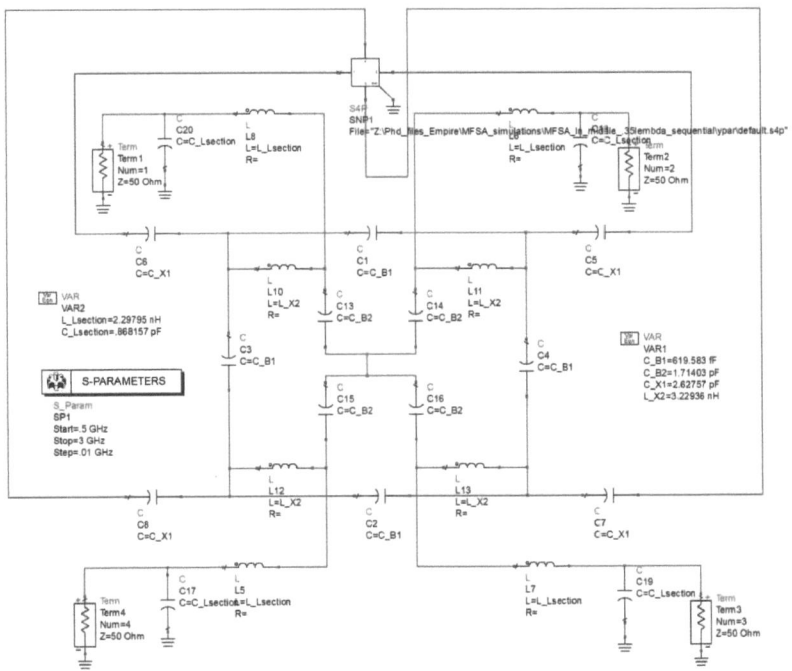

# APPENDIX E

**Layout of the Eigenmode Based Decoupling Network in MSL Realization for the MFSAA.**

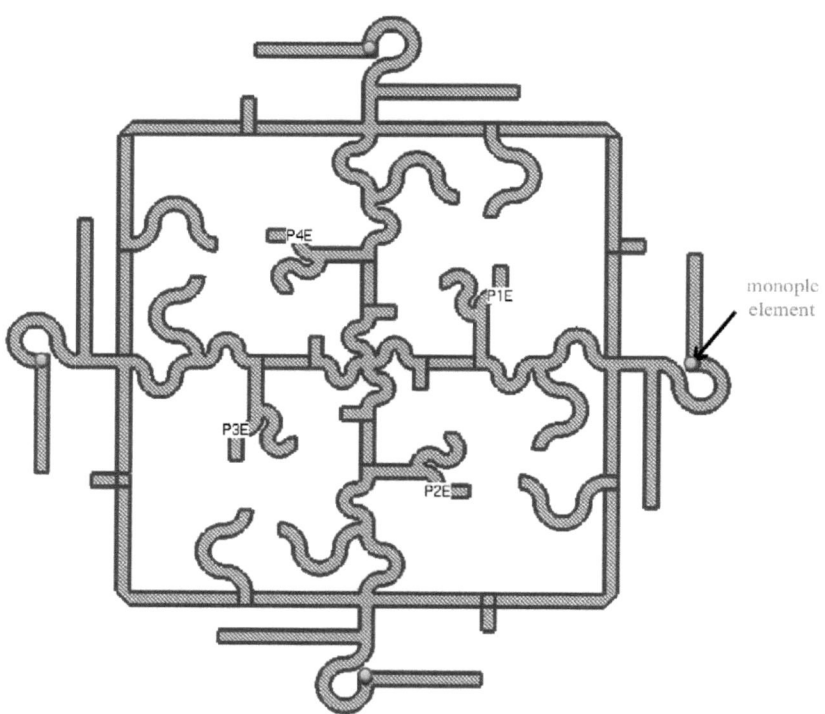

APPENDICES                                                                                        143

# APPENDIX F

**Layout of the Eigenmodal Feed Based Decoupling Network in MSL Realization for the MFSAA.**

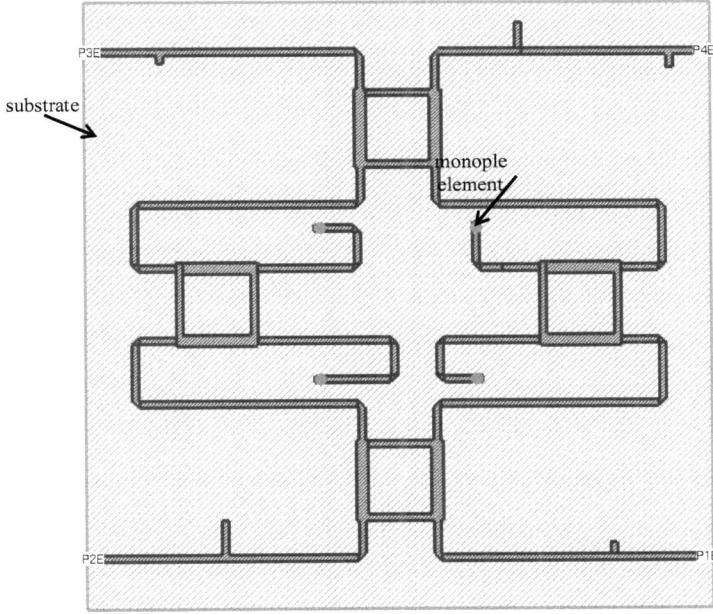

Printed by Books on Demand GmbH, Norderstedt / Germany